MALVI

Electronic

PRINCIPLES

SIXTH EDITION

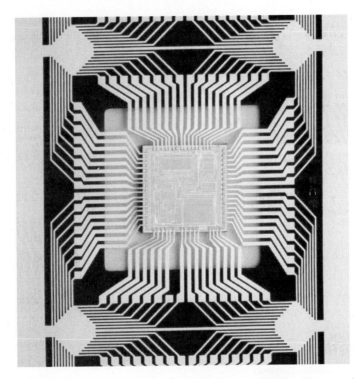

ALBERT PAUL MALVINO, PH.D., E.E.

Glencoe
McGraw-Hill

New York, New York   Columbus, Ohio   Woodland Hills, California   Peoria, Illinois

Cover photo: Dick Luria/FPG

Glencoe/McGraw-Hill

A Division of The McGraw·Hill Companies

Experiments for Electronic Principles, Sixth Edition

Send all inquiries to:
Glencoe/McGraw-Hill
8787 Orion Place
Columbus, OH 43240

ISBN 0-02-802834-1

Printed in the United States of America.

8 9   045   06 05 04

Contents

Preface

This laboratory manual contains 61 experiments to demonstrate the theory in *Electronic Principles*. Each experiment begins with a short discussion. Then, the Required Reading lists the sections of the textbook that should be read before you attempt to do the experiment. In the Procedure section, you will build and test one or more circuits. The Troubleshooting and Critical Thinking sections take you into more advanced areas. The Questions at the end of each experiment are a final test of what you have learned during the experiment.

Optional sections are included in many experiments. They are to be used at the discretion of the instructor. The Applications sections demonstrate how to use transducers such as buzzers, LEDs, microphones, motors, photoresistors, phototransistors, and speakers. The Additional Work sections include advanced experiments. The Computer section covers work with Electronics Workbench and the CD-ROM version of the textbook.

You will find the experiments in this lab manual to be instructive and interesting. The experiments verify and expand the theory presented in *Electronic Principles;* they make it come alive. When you have completed the experiments, you will have that rounded grasp of theory that can come only from practical experimentation.

Thanks to my special consultant, Richard Burchell, of ITT Technical Institute in Portland, Oregon. Again, his advice and suggestions were invaluable in deciding the content and format of this book.

Albert Paul Malvino

Voltage and Current Sources

An ideal or perfect voltage source produces an output voltage that is independent of the load resistance. A real voltage source, however, has a small internal resistance that produces an *IR* drop. As long as this internal resistance is much smaller than the load resistance, almost all the source voltage appears across the load. A stiff voltage source is one whose internal resistance is less than 1/100 of the load resistance. With a stiff voltage source, at least 99 percent of the source voltage appears across the load resistor.

A current source is different. It produces an output current that is independent of the load resistance. One way to build a current source is to use a source resistance that is much larger than the load resistance. An ideal current source has an infinite source resistance. A real current source has an extremely high source resistance. A stiff current source is one whose internal resistance is at least 100 times greater than the load resistance. With a stiff current source, at least 99 percent of the source current passes through the load resistor.

In this experiment you will build voltage and current sources, verifying the conditions necessary to get stiff sources. You can also troubleshoot and design sources.

Required Reading

Chapter 1 (Secs. 1-3 and 1-4) of *Electronic Principles*, 6th ed.

Equipment

1 power supply: adjustable to 10 V
6 ½-W resistors: 10 Ω, 47 Ω, 100 Ω, 470 Ω, 1 kΩ, 10 kΩ
1 VOM (analog or digital multimeter)

Procedure

VOLTAGE SOURCES

1. The circuit left of the *AB* terminals in Fig. 1-1 represents a voltage source and its internal resistance *R*. Before you measure any voltage or current, you should have an estimate of its value. Otherwise, you really don't know what you're doing. Look at Fig. 1-1 and estimate the load voltage for each value of *R* listed in Table 1-1. Record your rough estimates. Don't use a calculator to get these load voltages. Instead, mentally work out ballpark answers. All you're trying to do here is get into the habit of mentally estimating values before they are measured.

2. Connect the circuit in Fig. 1-1 using the values of *R* given in Table 1-1. Measure and adjust the source voltage to 10 V. For each *R* value, measure and record V_L.

CURRENT SOURCE

3. The circuit left of the *AB* terminals in Fig. 1-2 acts like a current source under certain conditions. Estimate and record the load current for each value of load resistance shown in Table 1-2.

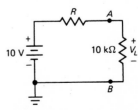

Figure 1-1

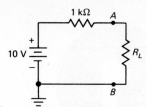

Figure 1-2

4. Connect the circuit of Fig. 1-2 using the R_L values given in Table 1-2. Measure and adjust the source voltage to 10 V. For each R_L value, measure and record I_L.

TROUBLESHOOTING

5. Connect the circuit of Fig. 1-1 with an R of 470 Ω. Connect a jumper wire between A and B. Measure the voltage across the load resistor and record your answer in Table 1-3.
6. Remove the jumper and open the load resistor. Measure the load voltage between the AB terminals and record in Table 1-3.

CRITICAL THINKING

7. Select an internal resistance R for the circuit of Fig. 1-1 to get a stiff voltage source for all load resistances greater than 10 kΩ. Connect the circuit of Fig. 1-1 using your design value of R. Measure the load voltage. Record the value of R and the load voltage in Table 1-4.
8. Select an internal resistance R for the circuit of Fig. 1-2 to get a stiff current source for all load resistances less than 100 Ω. Connect the circuit with your design value of R and a load resistance of 100 Ω. Measure the load current. Record the value of R and the load current in Table 1-4.

COMPUTER (OPTIONAL)

9. Use Electronics Workbench (EWB) or an equivalent circuit simulator to build the circuit of Fig. 1-1. Measure the load voltage for each source resistance shown in Table 1-1. Your measured values should be similar to those recorded in Table 1-1. (*Note:* To simulate properly, EWB needs a ground added to the negative battery terminal.)
10. Use EWB to build the circuit of Fig. 1-2. Measure the load current for each load resistance shown in Table 1-2 and compare each measurement to the recorded value.
11. Using EWB, construct the circuit you designed in Step 7. Verify the value of load voltage you recorded in Table 1-4.
12. Now, construct the circuit you designed in Step 8 and again use EWB to verify the value of load current recorded in Table 1-4.
13. If you are using the CD-ROM version of this book, click on the Assignments menu and select Chap. 1.

2

Data for Experiment 1

TABLE 1-1. VOLTAGE SOURCE

R	Estimated V_L	Measured V_L
0 Ω		
10 Ω		
100 Ω		
470 Ω		

TABLE 1-2. CURRENT SOURCE

R_L	Estimated I_L	Measured I_L
0 Ω		
10 Ω		
47 Ω		
100 Ω		

TABLE 1-3. TROUBLESHOOTING

Trouble	Measured V_L
Shorted load	
Open load	

TABLE 1-4. CRITICAL THINKING

Type	R	Measured Quantity
Voltage source		
Current source		

Questions for Experiment 1

1. The data of Table 1-1 prove that load voltage is: ( )
   (a) perfectly constant; **(b)** small; **(c)** heavily dependent on load resistance;
   (d) approximately constant.
2. When internal resistance R increases in Fig. 1-1, load voltage: ( )
   (a) increases slightly; **(b)** decreases slightly; **(c)** stays the same.
3. In Fig. 1-1, the voltage source is stiff when R is less than: ( )
   (a) 0 Ω; **(b)** 100 Ω; **(c)** 500 Ω; **(d)** 1 kΩ.
4. The circuit left of the AB terminals in Fig. 1-2 acts approximately like a current ( )
   source because the current values in Table 1-2:
   (a) increase slightly; **(b)** are almost constant; **(c)** decrease a great deal;
   (d) depend heavily on load resistance.
5. In Fig. 1-2, the circuit acts like a stiff current source as long as the load resistance is: ( )
   (a) less than 10 Ω; **(b)** large; **(c)** much larger than 1 kΩ; **(d)** greater
   than 1 kΩ.

6. Briefly explain the difference between a stiff voltage source and a stiff current source.

TROUBLESHOOTING

7. Explain why the load voltage with a shorted load is zero in Table 1-3. Consider using Ohm's law in your explanation.

8. Briefly explain why the load voltage with an open load is approximately equal to the source voltage in Table 1-3. Consider using Ohm's law and Kirchhoff's voltage law in your explanation.

CRITICAL THINKING

9. You are designing a current source that must appear stiff to all load resistances less than 10 kΩ. What is the minimum internal resistance your source can have? Explain why you selected this answer.

10. Optional: Instructor's question.

2

Thevenin's Theorem

The Thevenin voltage is the voltage that appears across the load terminals when you open the load resistor. The Thevenin voltage is also called the open-circuit or open-load voltage. The Thevenin resistance is the resistance between the load terminals with the load disconnected and all sources reduced to zero. This means replacing voltage sources by short circuits and current sources by open circuits.

In this experiment you will calculate the Thevenin voltage and resistance of a circuit. Then you will measure these quantities. Also included are troubleshooting and design options.

Required Reading

Chapter 1 (Sec. 1-5) of *Electronic Principles*, 6th ed.

Equipment

1 power supply: 15 V (adjustable)
7 ½-W resistors: 470 Ω, two 1 kΩ, two 2.2 kΩ, two
   4.7 kΩ
1 potentiometer: 5 kΩ
1 VOM (analog or digital multimeter)

Procedure

1. In Fig. 2-1a, calculate the Thevenin voltage V_{TH} and the Thevenin resistance R_{TH}. Record these values in Table 2-1.
2. With the Thevenin values just found, calculate the load voltage V_L across an R_L of 1 kΩ (see Fig. 2-1b). Record V_L in Table 2-2.
3. Also calculate the load voltage V_L for an R_L of 4.7 kΩ as shown in Fig. 2-1c. Record the calculated V_L in Table 2-2.
4. Connect the circuit of Fig. 2-1a, leaving out R_L.
5. Measure and adjust the source voltage of 15 V. Measure V_{TH} and record the value in Table 2-1.
6. Replace the 15-V source by a short circuit. Measure the resistance between the AB terminals using a convenient resistance range of the VOM. Record R_{TH} in Table 2-1. Now replace the short by the 15-V source.

7. Connect a load resistance R_L of 1 kΩ between the AB terminals of Fig. 2-1a. Measure and record load voltage V_L (Table 2-2).
8. Change the load resistance from 1 kΩ to 4.7 kΩ. Measure and record the new load voltage.
9. Find R_{TH} by the matched-load method; that is, use the potentiometer as a variable resistance between the AB terminals of Fig. 2-1a. Vary resistance until load voltage drops to half of the measured V_{TH}. Then disconnect the load resistance and measure its resistance with

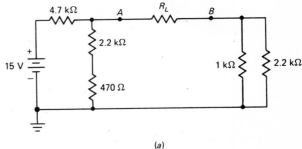

(a)

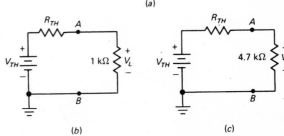

(b)   (c)

Figure 2-1

the VOM. This value should agree with R_{TH} found in Step 6.

TROUBLESHOOTING

10. Put a jumper wire across the 2.2-kΩ resistor, the one on the left side of Fig. 2-1a. Estimate the Thevenin voltage and Thevenin resistance for this trouble and record your rough estimates in Table 2-3. Measure the Thevenin voltage and Thevenin resistance (similar to Steps 5 and 6). Record the measured data in Table 2-3.

11. Remove the jumper wire and open the 2.2-kΩ resistor of Fig. 2-1a. Estimate and record the Thevenin quantities (Table 2-3). Measure and record the Thevenin quantities.

CRITICAL THINKING

12. Select the resistors for the unbalanced Wheatstone bridge of Fig. 2-2 to meet these specifications: $V_{TH} =$ 4.35 V and $R_{TH} = 3$ kΩ. Resistor values must be from those specified under the heading "Equipment." Record your design values in Table 2-4. Connect your circuit. Measure and record the Thevenin quantities.

13. Measure the output resistance R_{TH} of the signal generator or function generator that you will be using as a signal source for the rest of the experiments in this manual. Use the matched-load method as follows: First, adjust the open-circuit output voltage of the gen-

erator to exactly 1 V rms measured by the digital multimeter (DMM). Do not change the output voltage adjustment for the balance of this procedure. Then, connect a 1-kΩ potentiometer (connected as a rheostat) as a load resistor on the output of the generator in parallel with the DMM. Now, without changing the amplitude setting of the generator, adjust the potentiometer until the DMM reads exactly 0.5 V rms. Disconnect the DMM and the potentiometer from the generator and measure the resistance of the potentiometer. Record this value in Table 2-5 as the output resistance of the generator. It will be useful in later experiments.

COMPUTER (OPTIONAL)

14. Construct Fig. 2-2 in a computer simulation program similar to Electronics Workbench (EWB), with the design values that you specified in Step 12. Connect a dc voltmeter between points A and B instead of R_L. When power is applied to the circuit, you should read the value of V_{TH} within a reasonable tolerance of the measured value that you recorded in Table 2-4.

15. Now, replace the battery in your computer circuit with a short circuit (straight wire) and change the voltmeter to an ohmmeter. When power is applied, you should read the value of R_{TH} within a reasonable tolerance of the measured value that you recorded in Table 2-4.

16. If you are using the CD-ROM version of this book, click on the Assignments menu and select Chap. 1.

ADDITIONAL WORK (OPTIONAL)

17. Measure the output impedance of a signal generator or a function generator at several frequencies. Draw the Thevenin equivalent circuit for the generator. Graph z_{out} versus frequency.

18. Use an oscilloscope and a DMM to measure the sinusoidal output voltage of a signal generator or a function generator at various frequencies above 1 kHz. Notice how the measurements differ. One is a visual display, and the other is an rms reading.

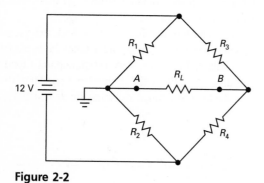

Figure 2-2

Data for Experiment 2

TABLE 2-1. THEVENIN VALUES

	V_{TH}	R_{TH}
Calculated		
Measured		

TABLE 2-2. LOAD VOLTAGES

	V_L for 1 kΩ	V_L for 4.7 kΩ
Calculated		
Measured		

TABLE 2-3. TROUBLESHOOTING

	Estimated		Measured	
	V_{TH}	R_{TH}	V_{TH}	R_{TH}
Shorted 2.2 kΩ				
Open 2.2 kΩ				

TABLE 2-4. CRITICAL THINKING

Design values: $R_1 =$ _____

$R_2 =$ _____

$R_3 =$ _____

$R_4 =$ _____

Measured values: $V_{TH} =$ _____

$R_{TH} =$ _____

TABLE 2-5.

Generator output resistance:

$R_{TH} =$ _____

Questions for Experiment 2

1. In this experiment you measured Thevenin voltage with: ( )
   (a) an ohmmeter; **(b)** the load disconnected; **(c)** the load in the circuit.
2. You first measured R_{TH} with a: ( )
   (a) voltmeter; **(b)** load; **(c)** shorted source.
3. You also measured R_{TH} by the matched-load method, which involves: ( )
   (a) an open voltage source; **(b)** a load that is open; **(c)** varying Thevenin resistance until it matches load resistance; **(d)** changing load resistance until load voltage drops to $V_{TH}/2$.

4. Discrepancies between calculated and measured values in Table 2-1 may be caused   ( )
   by:
   (a) instrument error;   (b) resistor tolerance;   (c) human error;   (d) all
   the foregoing.
5. If a black box puts out a constant voltage for all load resistances, the Thevenin re-   ( )
   sistance of this box approaches:
   (a) zero;   (b) infinity;   (c) load resistance.
6. Ideally, a voltmeter should have infinite resistance. Explain how a voltmeter with an input
   resistance of 100 kΩ will introduce a small error in Step 5 of the procedure.

TROUBLESHOOTING

7. Briefly explain why the Thevenin voltage and resistance are both lower when the 2.2-kΩ
   resistor is shorted.

8. Explain why V_{TH} and R_{TH} are higher when the 2.2-kΩ resistor is open.

CRITICAL THINKING

9. If you were manufacturing automobile batteries, would you try to produce a very low internal resistance or a very high internal resistance? Explain your reasoning.

10. Optional: Instructor's question.

8

Troubleshooting

A n open device always has zero current and unknown voltage. You have to figure out what the voltage is by looking at the rest of the circuit. On the other hand, a shorted device always has the zero voltage and unknown current. You have to figure out what the current is by looking at the rest of the circuit. In this experiment, you will insert troubles into a basic circuit. Then you will calculate and measure the voltages of the circuit.

Required Reading

Chapter 1 (Sec. 1-7) of *Electronic Principles*, 6th ed.

Equipment

1   power supply: 10 V
4   ½-W resistors: 1 kΩ, 2.2 kΩ, 3.9 kΩ, and 4.7 kΩ
1   VOM (analog or digital)

Procedure

1. Connect the circuit shown in Fig. 3-1.
2. Calculate the voltage between node A and ground. Record the value in Table 3-1 under "Circuit OK."
3. Calculate the voltage between node B and ground. Record the value.
4. Measure the voltages at A and B. Record these values in Table 3-1.
5. Open resistor R_1. Calculate the voltages at nodes A and B. Record these values in Table 3-1. Next, measure the voltages at nodes A and B. Record the values.
6. Repeat Step 5 for each of the remaining resistors listed in Table 3-1.

7. Short-circuit resistor R_1 by placing a jumper wire across it. Calculate and record the voltages in Table 3-1.
8. Repeat Step 7 for each of the remaining resistors in Table 3-1.

COMPUTER (OPTIONAL)

9. Repeat Steps 1 to 8 using EWB or an equivalent circuit simulator. Do not record values. The EWB measurements should be in reasonable agreement with the values recorded in Table 3-1.
10. If you are using the CD-ROM version of this book, click on the Assignments menu and select Chap. 1.

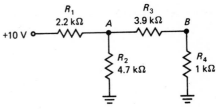

Figure 3-1

Data for Experiment 3

TABLE 3-1. TROUBLES AND VOLTAGES

	Calculated		Measured	
Trouble	V_A	V_B	V_A	V_B
Circuit OK				
R_1 open				
R_2 open				
R_3 open				
R_4 open				
R_1 shorted				
R_2 shorted				
R_3 shorted				
R_4 shorted				

Questions for Experiment 3

1. When R_1 is open in Fig. 3-1, V_A is approximately: ( )
    (a) 0;   **(b)** 1.06 V;   **(c)** 1.41 V;   **(d)** 6.81 V.

2. When R_2 is open in Fig. 3-1, V_B is approximately: ( )
    (a) 0;   **(b)** 1.06 V;   **(c)** 1.41 V;   **(d)** 6.81 V.

3. When R_3 is open in Fig. 3-1, V_A is approximately: ( )
    (a) 0;   **(b)** 1.06 V;   **(c)** 1.41 V;   **(d)** 6.81 V.

4. When R_4 is open in Fig. 3-1, V_B is approximately: ( )
    (a) 0;   **(b)** 1.06 V;   **(c)** 1.41 V;   **(d)** 6.81 V.

5. When R_1 is shorted in Fig. 3-1, V_A is approximately: ( )
    (a) 0;   **(b)** 2.04 V;   **(c)** 2.72 V;   **(d)** 10 V.

6. When R_2 is shorted in Fig. 3-1, V_B is approximately: ( )
    (a) 0;   **(b)** 2.04 V;   **(c)** 2.72 V;   **(d)** 10 V.

7. When R_3 is shorted in Fig. 3-1, V_A is approximately: ( )
    (a) 0;   **(b)** 2.04 V;   **(c)** 2.72 V;   **(d)** 10 V.

8. When R_4 is shorted in Fig. 3-1, V_B is approximately: ( )
    (a) 0;   **(b)** 1.06 V;   **(c)** 1.41 V;   **(d)** 6.81 V.

9. When R_3 is open in Fig. 3-1, V_B is approximately: ( )
    (a) 0;   **(b)** 1.06 V;   **(c)** 1.41 V;   **(d)** 4.92 V.

10. When R_4 is shorted in Fig. 3-1, V_A is approximately: ( )
    (a) 0;   **(b)** 1.06 V;   **(c)** 1.41 V;   **(d)** 4.92 V.

Semiconductor Diodes

An ideal diode acts like a switch that is closed when forward-biased and open when reverse-biased. One way to test a diode is with an analog ohmmeter. When the diode is forward-biased, the ohmmeter is measuring the forward resistance R_F. When the diode is reverse-biased, the ohmmeter is measuring the reverse resistance R_R. With silicon diodes, the ratio of R_R to R_F is more than 1000:1.

If you check a diode with a digital multimeter (DMM), another approach must be used. Most DMMs have a special position (marked with the diode symbol) for testing diodes. To check forward voltage, connect the red lead to the anode (unmarked), and the black lead to the cathode (marked with a colored band). A typical DMM should read 0.5 to 0.7 V for silicon diodes, 0.2 to 0.4 V for germanium diodes, and 1.4 to 2 V for LEDs. A reading near zero indicates a shorted diode, and an overrange indicates an open diode. When you reverse the leads, overrange should be displayed. A reading less than overrange indicates a very leaky diode.

Testing a diode with an ohmmeter or a DMM is an incomplete test because it checks only for major defects such as shorts, opens, or very leaky diodes. The ohmmeter cannot detect more subtle problems. For instance, a slightly leaky silicon diode may have a reverse resistance of only 10 kΩ, low enough to prevent it from working in many circuits. If you test this slightly leaky diode with a typical DMM, it will pass because the DMM will still indicate overrange when the diode is reverse-biased.

The point is this: Testing a diode with an ohmmeter or DMM is conclusive only when the diode fails the test. If it passes the test, the diode may still have some defects that prevent it from working in an ac circuit. Therefore, even though a diode may pass dc testing with an ohmmeter, it must still be checked in a working circuit before you can be sure that it is all right.

Required Reading

Chapter 2 and Chap. 3 (Secs. 3-1 and 3-2) of *Electronic Principles*, 6th ed.

Equipment

1   power supply: adjustable to 10 V
1   VOM (analog multimeter)
1   DMM (digital multimeter)
Diodes: Four 1N4001
1   ½-W resistor: 1 kΩ

Procedure

1. In this part of the experiment, you will be measuring the forward and reverse resistance of a diode.
2. Do you expect the forward resistance of a diode to be low or high? Record your answer in Table 4-1.
3. Repeat Step 2 for the reverse resistance of a diode.
4. With an analog ohmmeter set to the X1K range, measure the resistance of a 1N4001 in either direction. (The polarity of the red and black leads of the ohmmeter doesn't matter because the red lead may be positive or negative with analog ohmmeters.) Then reverse the leads and measure in the other direction. You

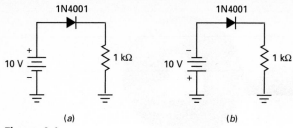

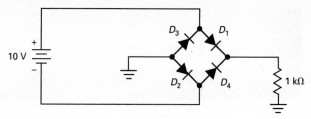

(a)

(b)

Figure 4-1

Figure 4-3

should get a low reading one way and a high reading the other way. (The exact values don't matter because the resistance of the diode will depend on the ohm-meter range.)

5. Record your readings in Table 4-1 under "Measured 1." If the reverse resistance is too high to read, record "open."

6. Repeat Steps 4 and 5 for two more diodes.

7. Next, use a DMM to test a diode as follows: Select the special diode position (marked with a diode symbol). Connect the red lead to the anode and the black lead to the cathode of a 1N4001. With DMMs, the red and black leads are polarized because the red lead is always positive on the special diode position. The DMM will forward-bias the diode and display the voltage across it. Record the voltage in Table 4-2 under "Measured 1."

8. The reading you got in the preceding step depends on the DMM used. Since a 1N4001 is a silicon diode, the reading should be from 0.5 to 0.7 V, depending on how much current the DMM produces in the diode.

9. Now, reverse the leads and you should get an over-range display (typically shown as OL). Record the reading in Table 4-2 (use OL if it is an overrange).

10. Repeat Steps 7 to 9 for two more diodes.

11. In Fig. 4-1a, calculate the voltage across the diode and across the load resistor. Record your calculated values in Table 4-3.

12. Build the circuit of Fig. 4-1a. Measure and record the diode and load voltages in Table 4-3.

13. Repeat Steps 11 and 12 for two more diodes.

14. In Fig. 4-1b, calculate the voltage across the diode and across the load resistor. Record your calculated values in Table 4-4.

15. Build the circuit of Fig. 4-1b. Measure and record the diode and load voltages in Table 4-4.

16. Repeat Step 15 for two more diodes.

17. Figure 4-2 shows a diode circuit. Is diode D_1 on or off? Record your answer in Table 4-5 under "D_1 normal."

18. Repeat Step 17 for the remaining diodes shown in Fig. 4-2.

19. Assume that the polarity of the battery is reversed in Fig. 4-2. Determine whether each diode is on or off. Record your answers in Table 4-5 for the reverse condition.

20. In Fig. 4-3, determine whether each diode is on or off. Record your answers in Table 4-6.

21. Assume that the polarity of the battery is reversed in Fig. 4-3. Record the on-off condition of each diode in Table 4-6.

22. Calculate the voltage across each diode in Fig. 4-3. Also calculate the load voltage. Record all values in Table 4-7.

23. Build the circuit of Fig. 4-3. Measure and record all voltages listed in Table 4-7.

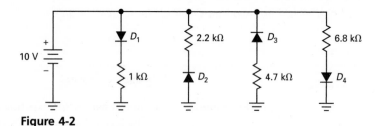

Figure 4-2

Data for Experiment 4

TABLE 4-1. OHMMETER TESTING

	Expected	Measured 1	Measured 2	Measured 3
R_F				
R_R				

TABLE 4-2. DMM TESTING

	Measured 1	Measured 2	Measured 3
Forward			
Reverse			

TABLE 4-3. DATA FOR FORWARD BIAS

	Calculated		Measured	
	V_D	V_L	V_D	V_L
Diode 1				
Diode 2				
Diode 3				

TABLE 4-4. DATA FOR REVERSE BIAS

	Calculated		Measured	
	V_D	V_L	V_D	V_L
Diode 1				
Diode 2				
Diode 3				

TABLE 4-5. DIODE CONDUCTION

	D_1	D_2	D_3	D_4
Normal				
Reversed				

TABLE 4-6. DIODE CONDUCTION

	D_1	D_2	D_3	D_4
Normal				
Reversed				

TABLE 4-7. DIODE AND LOAD VOLTAGES

	V_{D1}	V_{D2}	V_{D3}	V_{D4}	V_L
Calculated					
Measured					

Questions for Experiment 4

1. A forward-biased diode ideally appears as:
(**a**) a closed switch;   (**b**) an open switch;   (**c**) a high resistance;   (**d**) an   ( )
insulator.

2. A reverse-biased diode ideally appears as:
(**a**) a closed switch;   (**b**) an open switch;   (**c**) a low resistance;   (**d**) a   ( )
conductor.

3. When reverse-biased, a diode:
(**a**) appears shorted;   (**b**) has a low resistance;   (**c**) has 0 V across it;   ( )
(**d**) appears open.

4. When a diode is tested with a DMM, the indication with reverse bias is normally:
(**a**) 0.5 V;   (**b**) 0.7 V;   (**c**) an overrange;   (**d**) low.   ( )

5. The diode of Fig. 4-1*b* is:
(**a**) conducting heavily;   (**b**) reverse-biased;   (**c**) forward-biased;   (**d**) on.   ( )

6. In Fig. 4-2, diode D_3 is:
(**a**) conducting heavily;   (**b**) reverse-biased;   (**c**) forward-biased;   (**d**) on.   ( )

7. In Fig. 4-3, diode D_2 is:
(**a**) not conducting;   (**b**) reverse-biased;   (**c**) forward-biased;   (**d**) off.   ( )

8. If a diode in Fig. 4-3 has a voltage of 0.7 V across it when conducting, the load
voltage will be: ( )
(**a**) 0;   (**b**) 8.6 V;   (**c**) 9.3 V;   (**d**) 10 V.

9. If a diode D_1 opens in Fig. 4-3, the load voltage will be:
(**a**) 0;   (**b**) 8.6 V;   (**c**) 9.3 V;   (**d**) 10 V.   ( )

10. If a diode in Fig. 4-3 has a voltage of 0.7 V across it when conducting, the voltage across D_4 will be: ( )
(**a**) 0;   (**b**) 8.6 V;   (**c**) 9.3 V;   (**d**) 10 V.

The Diode Curve

A resistor is a linear device because its voltage and current are proportional in either direction. A diode, on the other hand, is a nonlinear device because its current and voltage are not proportional. Furthermore, a diode is a unilateral device because it conducts well only in the forward direction. As a guide, a small-signal silicon diode has a dc reverse/forward resistance ratio of more than 1000 : 1. In this experiment you will measure diode currents and voltages for both forward and reverse bias. This will allow you to draw the diode curve. Also included are troubleshooting, design, and computer operations.

Required Reading

Chapter 2 (Secs. 2-8 to 2-11) and Chap. 3 (Sec. 3-1) of *Electronic Principles*, 6th ed.

Equipment

1 power supply: adjustable from approximately 0 to 15 V
1 diode: 1N4148 or 1N914 (small-signal diode)
3 ½-W resistors: 220 Ω, 1 kΩ, 100 kΩ
1 VOM (analog or digital multimeter)
1 milliammeter or another VOM if available
Graph paper, rectangular coordinates

Procedure

OHMMETER TEST

1. Using the VOM as an ohmmeter, measure a 1N4148's dc forward resistance and reverse resistance on one of the middle resistance ranges. If the diode is all right, you should have a reverse/forward ratio greater than 1000 : 1.

DIODE DATA

2. Connect the circuit of Fig. 5-1 using a current-limiting resistor of 1 kΩ. For each source voltage listed in Table 5-1, measure and record the diode voltage V and the diode current I.

3. Calculate and record the dc forward resistance of the diode for each current of Table 5-1.

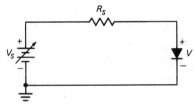

Figure 5-1

4. Reverse the source voltage in Fig. 5-1. For each source voltage of Table 5-2, measure and record the diode voltage V and the diode current I.

5. Calculate and record the dc reverse resistance of the diode for each source voltage of Table 5-2.

6. Graph the data of Tables 5-1 and 5-2 to get a diode curve (I versus V).

7. The foregoing steps prove that the diode conducts easily in the forward direction and poorly in the reverse direction. It's like a one-way conductor. With this in mind, estimate the diode current in Fig. 5-2a and b. Record your ballpark estimates in Table 5-3.

8. Connect the circuit of Fig. 5-2a (forward bias). Measure and record the diode current in Table 5-3.

9. Connect the circuit of Fig. 5-2b (reverse bias). Measure and record the diode current.

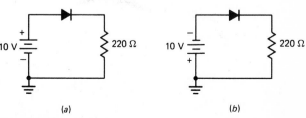

Figure 5-2

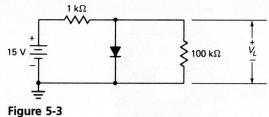

Figure 5-3

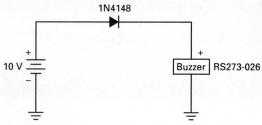

Figure 5-5

TROUBLESHOOTING

10. Connect the circuit of Fig. 5-3. Estimate the load voltage V_L and record in Table 5-4. Then measure and record V_L.

11. Short the diode with a jumper wire. Estimate V_L for this condition and record in Table 5-4. Measure and record V_L.

12. Remove the jumper wire. Disconnect one end of the diode. Estimate V_L and record. Next, measure and record V_L.

CRITICAL THINKING

13. Select a source voltage and a current-limiting resistance to produce 10 mA in Fig. 5-1a. (Use the resistors from earlier parts of the experiment.) Connect your circuit and measure the current. Record your values of V_S and R_S, along with the measured I in Table 5-5.

APPLICATION (OPTIONAL)

14. Connect the circuit of Fig. 5-4 using a magnetic buzzer equivalent to the Radio Shack RS273-026, which has

the following specifications: 6- to 16-V dc input and 10-mA load current at 12 V dc. Notice that the buzzer is polarized, so you must connect the positive source terminal to the red lead and the negative source terminal to the black lead. You should hear a buzz.

15. Look at Fig. 5-5. Should the buzzer work? Connect the circuit of Fig. 5-5 to verify your answer.

16. Repeat Step 15 for the circuit of Fig. 5-6.

COMPUTER (OPTIONAL)

17. Repeat Steps 1 to 13 using EWB or an equivalent circuit simulator. Do not record any new values. But make sure that you get reasonable agreement between the EWB measurements and the values recorded earlier.

18. If you are using the CD-ROM version of this book, click on the Assignments menu and select Chap. 3.

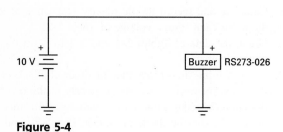

Figure 5-4

Figure 5-6

Data for Experiment 5

TABLE 5-1. FORWARD BIAS

V_S	V	I	R
0			(no entry)
0.5 V			
1 V			
2 V			
4 V			
6 V			
8 V			
10 V			
15 V			

TABLE 5-2. REVERSE BIAS

V_S	V	I	R
−1 V			
−5 V			
−10 V			
−15 V			

TABLE 5-3. DIODE CONDUCTION

	Estimated I	Measured I
Fig. 5-2a		
Fig. 5-2b		

TABLE 5-4. TROUBLESHOOTING

	Estimated V_L	Measured V_L
Normal diode		
Shorted diode		
Open diode		

TABLE 5-5. CRITICAL THINKING

$V_S =$ _____ $R_S =$ _____ $I =$ _____

Questions for Experiment 5

1. In this experiment the knee or offset voltage is closest to: ( )
   (a) 0.3 V;   (b) 0.7 V;   (c) 1 V;   (d) 1.2 V.
2. With forward bias, the dc resistance decreases when: ( )
   (a) current increases;   (b) diode decreases;   (c) the ratio V/I increases;
   (d) the ratio I/V decreases.
3. A diode acts like a high resistance when: ( )
   (a) its current is large;   (b) forward-biased;   (c) reverse-biased;
   (d) shorted.
4. Which of the following approximately describes the diode curve above the forward   ( )
   knee?
   (a) it becomes horizontal;   (b) voltage increases rapidly;   (c) current increases rapidly;   (d) dc resistance increases rapidly.
5. Which of the following describes the diode curve in the reverse direction?   ( )
   (a) ratio I/V is high;   (b) it becomes vertical below breakdown;   (c) dc resistance is low;   (d) current is approximately zero below breakdown.
6. Briefly describe how a diode differs from an ordinary resistor:

TROUBLESHOOTING

7. Why is the load voltage around 0.7 V in Fig. 5-3 when the diode is OK?

8. Why is the load voltage slightly less than 15 V when the diode is open in Fig. 5-3?

CRITICAL THINKING

9. If you are trying to set up a fixed current through any diode, is it better to use a low or a high source voltage? Explain your reasoning.

10. Optional: Instructor's question.

Diode Approximations

In the ideal or first approximation, a diode acts like a closed switch when forward-biased and an open switch when reversed-biased. In the second approximation, we include the knee voltage of the diode when it is forward-biased. This means assuming 0.7 V across a conducting silicon diode (0.3 V for germanium). The third approximation includes the knee voltage and the bulk resistance; because of this, the voltage across a conducting diode increases as the diode current increases. In troubleshooting and design, the second approximation is usually adequate.

In this experiment, you will work with the three diode approximations. Also included are troubleshooting, design, and computer exercises.

Required Reading

Chapter 3 (Secs. 3-2 to 3-5) of *Electronic Principles,* 6th ed.

Equipment

1   power supply: adjustable from approximately 0 to 15 V
1   diode: 1N4148 or 1N914
3   ½-W resistors: two 220 Ω, 470 Ω
1   VOM (analog or digital multimeter)

Procedure

1. Connect the circuit of Fig. 6-1a. Adjust the source to set up a current of 10 mA through the diode. Estimate the diode voltage V and record in Table 6-1.

2. Measure the diode voltage V and record in Table 6-1.
3. Adjust the source to get 50 mA. Estimate the diode voltage and record in Table 6-1. Measure and record diode voltage V.
4. In this experiment, we will let the knee voltage be the measured diode voltage for a diode current of 10 mA. Record the knee voltage in Table 6-2. (It should be in the vicinity of 0.7 V.)
5. Calculate the bulk resistance using

$$r_B = \frac{\Delta V}{\Delta I}$$

where ΔV and ΔI are the changes in measured voltage and current in Table 6-1. Record r_B in Table 6-2.
6. Calculate the diode current in Fig. 6-1b as follows: thevenize the circuit left of the AB terminals. Then calculate the diode current with the ideal, second, and third approximations (use the V_{knee} and r_B of Table 6-2). Record your answers in Table 6-3.
7. Connect the circuit of Fig. 6-1b. Measure and record the diode current (Table 6-3).

TROUBLESHOOTING

8. Estimate the diode current in Fig. 6-1b for each of these conditions: 470 Ω shorted and open. Record your rough estimates in Table 6-4.
9. Measure and record the diode current in the circuit of Fig. 6-1b with the 470-Ω resistor shorted and open.

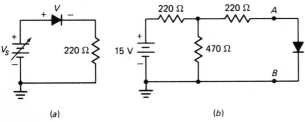

(a)   (b)

Figure 6-1

CRITICAL THINKING

10. Using the second approximation in Fig. 6-2, select values for resistors and source voltage to produce a diode current of approximately 8.9 mA. (Use the same resistance values as in Fig. 6-1*b*, although you can move the values.) Connect the circuit using your design values and measure the diode current. Record all data listed in Table 6-5.

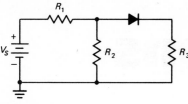

Figure 6-2

COMPUTER (OPTIONAL)

11. Repeat Steps 1 to 10 using EWB or an equivalent circuit simulator. Do not record any new values. But make sure that you get reasonable agreement between the EWB measurements and the values recorded earlier.

12. If you are using the CD-ROM version of this book, click on the Assignments menu and select Chap. 3.

ADDITIONAL WORK (OPTIONAL)

13. Repeat Steps 1 to 7 for a germanium diode such as the 1N34A. Record your data on a separate piece of paper and then compare the values to those recorded for a silicon diode. What did you learn?

Data for Experiment 6

TABLE 6-1. TWO POINTS ON THE FORWARD CURVE

I	Estimated V	Measured V
10 mA		
50 mA		

TABLE 6-2. DIODE VALUES

$V_{knee} =$	
$r_B =$	

TABLE 6-3. DIODE CURRENT

Ideal $I =$	
Second $I =$	
Third $I =$	
Measured $I =$	

TABLE 6-4. TROUBLESHOOTING

	Estimated I	Measured I
Shorted 470 Ω		
Open 470 Ω		

TABLE 6-5. CRITICAL THINKING

Design	
$V_S =$	
$R_1 =$	
$R_2 =$	
$R_3 =$	
I (diode) $=$	

Questions for Experiment 6

1. In this experiment, knee voltage is the diode voltage that: ( )
   (a) equals 0.3 V; (b) equals 0.7 V; (c) corresponds to 10 mA; (d) corresponds to 50 mA.
2. Bulk resistance is: ( )
   (a) diode voltage divided by current; (b) the ratio of voltage difference to current difference above the knee; (c) the same as the dc resistance of the diode; (d) none of the foregoing.

3. The dc resistance of a silicon diode for a current of 10 mA is closest to: ( )
   (a) 2.5 Ω;   (b) 10 Ω;   (c) 70 Ω;   (d) 1 kΩ.
4. In Fig. 6-1b, the power dissipated by the diode is closest to: ( )
   (a) 0;   (b) 1.5 mW;   (c) 15 mW;   (d) 150 mW.
5. Suppose the diode of Fig. 6-1b has an $I_{F(max)}$ of 500 mA. To avoid diode damage, ( )
   the source voltage can be no more than
   (a) 15 V;   (b) 50 V;   (c) 185 V;   (d) 272 V.
6. The steeper the diode curve, the smaller the bulk resistance. Explain why this is true.

TROUBLESHOOTING

7. Explain why there is no diode current when the 470-Ω resistor is shorted in Fig. 6-1b.

8. Why does the diode current increase with an open 470-Ω resistor in Fig. 6-1b?

CRITICAL THINKING

9. How many designs are possible in Step 10 of the Procedure? ( )
   (a) 1;   (b) 2;   (c) 3;   (d) 4.
10. Optional: Instructor's question.

Rectifier Circuits

The three basic rectifier circuits are the half-wave, the full-wave, and the bridge. The ripple frequency of a half-wave rectifier is equal to the input frequency, whereas the ripple frequency of a full-wave or bridge rectifier is equal to twice the input frequency. For a given transformer, the unfiltered output of the half-wave and full-wave rectifiers ideally has a dc value of slightly less than half the rms secondary voltage (45 percent), and the unfiltered output of a bridge rectifier is slightly less than the rms secondary voltage (90 percent).

In this experiment you will build all three types of rectifiers and measure their input-output characteristics. Be especially careful in this experiment when connecting the transformer to line voltage. The transformer should have a fused line cord with all primary connections insulated to avoid electrical shock.

Required Reading

Chapter 4 (Secs. 4-1 to 4-4) of *Electronic Principles*, 6th ed.

Equipment

1   transformer, 12.6 V ac center-tapped (Triad F-25X or equivalent) with fused line cord
4   silicon diodes: 1N4001 (or equivalent)
1   ½-W resistor: 1 kΩ
1   VOM (analog or digital multimeter)
1   oscilloscope

Procedure

HALF-WAVE RECTIFIER

1. In Fig. 7-1a, the rms secondary output voltage is a nominal 12.6 V ac. Calculate the peak output voltage across the 1-kΩ load resistor. Also calculate the dc output voltage and ripple frequency. Record your calculations in Table 7-1.
2. Connect the half-wave rectifier shown in Fig. 7-1a.
3. Measure the rms voltage across the secondary winding and record in Table 7-1.
4. Measure and record the dc load voltage.
5. Use an oscilloscope to look at the rectified voltage across the 1-kΩ load resistor. Record the peak voltage

of the half-wave signal. Next, measure the period of the rectified output. Calculate the ripple frequency and record the result in Table 7-1.

FULL-WAVE RECTIFIER

6. In Fig. 7-1b, calculate and record the quantities listed in Table 7-2.
7. Connect the center-tap rectifier of Fig. 7-1b.
8. Measure and record the quantities listed in Table 7-2.

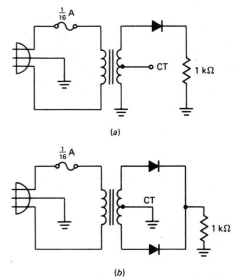

(a)

(b)

Figure 7-1

25

BRIDGE RECTIFIER

9. In Fig. 7-2, calculate the quantities listed in Table 7-3.
10. Connect the bridge rectifier of Fig. 7-2.
11. Measure and record the quantities listed in Table 7-3.

TROUBLESHOOTING

12. Assume that one of the diodes is open in the bridge rectifier. Calculate and record the dc output voltage and ripple frequency in Table 7-4.
13. Open one of the diodes. Measure and record the dc output voltage and ripple frequency. Restore the diode to a normal connection.
14. Assume that half of the secondary winding to the bridge rectifier is shorted (between the center tap and either end). Calculate and record the dc output voltage and ripple frequency in Table 7-4.
15. Simulate the foregoing short by disconnecting either end of the secondary and connecting the center tap in its place. Measure and record the dc output voltage and ripple frequency.

CRITICAL THINKING

16. Figure out how to modify the bridge rectifier of Fig. 7-2 to meet the following specifications: dc load voltage is approximately 5.67 V, and dc load current is approximately 20 mA. (You need to select a new load resistor.)
17. Get the required load resistor and connect the modified circuit. Measure and record all the quantities listed in Table 7-5.

APPLICATION (OPTIONAL)

18. The unfiltered output of a half-wave rectifier can be used to drive a motor. The motor will respond to

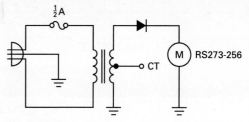

Figure 7-3

the dc or average value of the rectified voltage. As a demonstration, connect the circuit of Fig. 7-3 using a dc motor such as the Radio Shack RS273-256. This dc motor has the following specifications: 9- to 18-V dc input with a full-load current of 1.98 A and a no-load current of 300 mA. You will be running the motor under no-load conditions since the motor is not connected to a mechanical load. After the circuit is connected, you should be able to hear the motor running.

19. Reverse the diode as shown in Fig. 7-4. The motor will again turn, but this time in the opposite direction.
20. If a variac (variable line transformer) is available, you can use it to vary the line voltage to the circuit of Figs. 7-3 and 7-4. This will change the speed of the motor.

COMPUTER (OPTIONAL)

21. Repeat Steps 1 to 13 using EWB or an equivalent circuit simulator. Do not record any new values. But make sure that you get reasonable agreement between the EWB measurements and the values recorded earlier.
22. If you are using the CD-ROM version of this book, click on the Assignments menu and select Chap. 4.

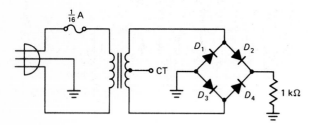

Figure 7-2

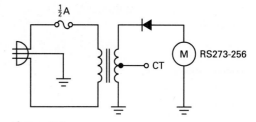

Figure 7-4

Data for Experiment 7

TABLE 7-1. HALF-WAVE RECTIFIER

	Calculated	Measured
RMS secondary voltage	12.6 V	
Peak output voltage		
DC output voltage		
Ripple frequency		

TABLE 7-2. FULL-WAVE RECTIFIER

	Calculated	Measured
RMS secondary voltage	12.6 V	
Peak output voltage		
DC output voltage		
Ripple frequency		

TABLE 7-3. BRIDGE RECTIFIER

	Calculated	Measured
RMS secondary voltage	12.6 V	
Peak output voltage		
DC output voltage		
Ripple frequency		

TABLE 7-4. TROUBLESHOOTING

	Calculated		Measured	
	V_{dc}	f_{out}	V_{dc}	f_{out}
Diode open				
Half-secondary short				

TABLE 7-5. CRITICAL THINKING

	Calculated	Measured
RMS secondary voltage	6.3 V	
Peak output voltage		
DC load voltage		
DC load current		
Ripple frequency		
Load resistance		

Questions for Experiment 7

1. To measure the rms secondary voltage, it is best to use: ( )
   (a) an oscilloscope; **(b)** an ammeter; **(c)** a voltmeter with the common lead grounded; **(d)** a floating VOM.
2. With the full-wave rectifier of this experiment, the dc load voltage was closest to: ( )
   (a) 1 V; **(b)** 3 V; **(c)** 6 V; **(d)** 12 V.
3. The dc load voltage out of the bridge rectifier compared with the full-wave rectifier was approximately: ( )
   (a) half as large; **(b)** the same; **(c)** twice as large; **(d)** 60 Hz.
4. Of the three rectifiers tested, the one with the largest dc output was: ( )
   (a) half-wave; **(b)** full-wave; **(c)** bridge; **(d)** no answer.
5. The unfiltered dc output voltage from a bridge rectifier is ideally what percent of the rms secondary voltage: ( )
   (a) 31.8; **(b)** 45; **(c)** 63.6; **(d)** 90.
6. Explain why the bridge rectifier is the most widely used of the three types.

TROUBLESHOOTING

7. Explain why the dc output voltage and ripple frequency of a bridge rectifier drop in half when any diode opens.

8. If any diode in a bridge rectifier is shorted for any reason (solder bridge, fused diode, etc.), what will happen to the other diodes when power is applied? Explain your answer briefly.

CRITICAL THINKING

9. Briefly explain what you did in your design and why you did it.

10. Optional: Instructor's question.

8

The Capacitor-Input Filter

By connecting the output of a bridge rectifier to a capacitor-input filter, we can produce a dc load voltage that is approximately constant. Ideally, the filtered dc output voltage equals the peak secondary voltage. To a better approximation, the dc voltage is typically 90 to 95 percent of the peak secondary voltage with a peak-to-peak ripple of about 10 percent.

In this experiment you will connect a bridge rectifier to a capacitor-input filter. By changing load resistors and filter capacitors, you will verify the basic relations discussed in the textbook. Be especially careful in this experiment when connecting the transformer to line voltage. The transformer should have a fused line cord with all primary connections insulated to avoid electrical shock.

Required Reading

Chapter 4 (Secs. 4-6 to 4-9) of *Electronic Principles*, 6th ed.

Equipment

1   transformer: 12.6 V ac center-tapped (Triad F-25X or equivalent) with fused line cord
4   silicon diodes: 1N4001 (or equivalent)
2   ½-W resistors: 1 kΩ, 10 kΩ
2   capacitors: 47 μF and 470 μF (25-V rating or better)
1   VOM (analog or digital multimeter)
1   oscilloscope

Procedure

1. Measure the resistance of the primary and secondary windings. Record in Table 8-1.
2. In Fig. 8-1, assume the rms secondary voltage is 12.6 V. Also assume $R_L = 1$ kΩ and $C = 47$ μF. Calculate and record the quantities listed in Table 8-2.
3. Build the circuit of Fig. 8-1 with $R_L = 1$ kΩ and $C = 47$ μF.
4. Measure and record all the quantities listed in Table 8-2.
5. Repeat Steps 2 through 4 for $R_L = 1$ kΩ and $C = 470$ μF. Use Table 8-3.
6. Repeat Steps 2 through 4 for $R_L = 10$ kΩ and $C = 470$ μF. Use Table 8-4.

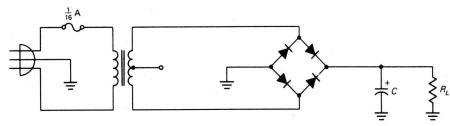

Figure 8-1

TROUBLESHOOTING

7. Assume that one of the diodes is open in Fig. 8-1, with $R_L = 1$ kΩ and $C = 470$ μF. Calculate the dc load voltage, ripple frequency, and peak-to-peak ripple. Record your results in Table 8-5.
8. Connect the foregoing circuit with one of the diodes open. Measure and record the quantities of Table 8-5.
9. Assume that the filter capacitor is open in Fig. 8-1, with $R_L = 1$ kΩ and $C = 470$ μF. Calculate and record the quantities listed in Table 8-5 for this trouble.
10. Connect the circuit of Fig. 8-1 with an open filter capacitor. Measure and record the remaining quantities of Table 8-5.

CRITICAL THINKING

11. Select a filter capacitor for the circuit of Fig. 8-1 to get a peak-to-peak ripple of about 10 percent of load voltage for an R_L of 3.9 kΩ. Calculate and record the quantities of Table 8-6.
12. Connect your circuit. Measure and record the quantities of Table 8-6.

APPLICATION (OPTIONAL)

13. Connect the circuit of Fig. 8-2 using a magnetic buzzer equivalent to the Radio Shack RS273-026, which has the following specifications: 6- to 16-V dc input and 10-mA load current at 12 V dc. The buzzer is polarized, so you must connect the positive source terminal to the red lead and the negative source terminal to the black lead. You should hear a buzz.

COMPUTER (OPTIONAL)

14. Repeat Steps 1 to 12 using EWB or an equivalent circuit simulator. Do not record any new values. But make sure that you get reasonable agreement between the EWB measurements and the values recorded earlier.
15. If you are using the CD-ROM version of this book, click on the Assignments menu and select Chap. 4.

ADDITIONAL WORK (OPTIONAL)

16. Connect a half-wave rectifier with a capacitor-input filter using a filter capacitance of 470 μF and a load resistance of 1 kΩ. Measure the load voltage and peak-to-peak ripple. Record the values.
17. Repeat Step 16 for a full-wave center-tap rectifier.
18. Compare the load voltage and peak-to-peak ripple of Step 16 with Step 17. What did you learn?

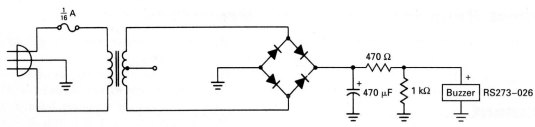

Figure 8-2

Data for Experiment 8

TABLE 8-1. TRANSFORMER RESISTANCES

$R_{pri} =$ _____

$R_{sec} =$ _____

TABLE 8-2. $R_L = 1$ kΩ AND $C = 47$ μF

	Calculated	Measured
RMS secondary voltage	12.6 V	
Peak output voltage		
DC output voltage		
DC load current		
Ripple frequency		
Peak-to-peak ripple		

TABLE 8-3. $R_L = 1$ kΩ AND $C = 470$ μF

	Calculated	Measured
RMS secondary voltage	12.6 V	
Peak output voltage		
DC output voltage		
DC load current		
Ripple frequency		
Peak-to-peak ripple		

TABLE 8-4. $R_L = 10$ kΩ AND $C = 470$ μF

	Calculated	Measured
RMS secondary voltage	12.6 V	
Peak output voltage		
DC output voltage		
DC load current		
Ripple frequency		
Peak-to-peak ripple		

TABLE 8-5. TROUBLESHOOTING

	Calculated				Measured		
	V_{dc}	f_{out}	V_{rip}		V_{dc}	f_{out}	V_{rip}
Open diode							
Open capacitor							

TABLE 8-6. CRITICAL THINKING

	Calculated	Measured
RMS secondary voltage		
Peak output voltage		
DC load voltage		
DC load current		
Ripple frequency		
Peak-to-peak ripple		

Questions for Experiment 8

1. In this experiment the dc output voltage from the capacitor-input filter was approximately equal to: ( )
   (a) peak primary voltage; (b) peak secondary voltage; (c) rms primary voltage; (d) rms secondary voltage.

2. The peak-to-peak ripple decreases when the: ( )
   (a) load resistance decreases; (b) filter capacitor decreases; (c) ripple frequency decreases; (d) filter capacitor increases.

3. When the filter capacitor increases, the peak-to-peak ripple: ( )
   (a) equals the secondary voltage; (b) remains constant; (c) increases; (d) decreases.

4. For normal operation, the ripple frequency is: ( )
   (a) 0; (b) 60 Hz; (c) 120 Hz; (d) 240 Hz.

5. When the load resistance increases, the peak-to-peak ripple: ( )
   (a) decreases; (b) stays the same; (c) increases; (d) none of the foregoing.

6. Briefly explain how a capacitor-input filter works.

TROUBLESHOOTING

7. When any diode opens, the circuit of Fig. 8-1 becomes a capacitor-input filter driven by a: ( )
   (a) half-wave rectifier; (b) full-wave rectifier; (c) bridge rectifier; (d) unilateral converter.

8. Briefly explain what happens to the circuit of Fig. 8-1 when the filter capacitor opens.

CRITICAL THINKING

9. What size of capacitor did you use in your design? Why did you select this size?

10. The turns ratio of the transformer is approximately 9:1. Using the transformer resistances in Table 8-1, calculate the minimum Thevenin resistance that is facing the filter capacitor. Ignore the bulk resistance of the diode and use only the primary and secondary winding resistance in your calculations.

Limiters and Peak Detectors

A positive limiter clips off positive parts of the input signal, and a negative limiter clips negative parts. In a biased limiter, the clipping level is selectable. With a combination limiter, positive and negative parts of the signal are removed. A diode clamp is an alternative name for a limiter. Often, a diode clamp is used to protect a load from excessively high input voltages.

In this experiment, you will connect different limiters. You will also experiment with a peak detector, a variation of the rectifier circuits discussed earlier. A peak detector produces a dc output voltage approximately equal to the peak voltage of the input signal.

Required Reading

Chapter 4 (Sec. 4-10) of *Electronic Principles*, 6th ed.

Equipment

1   audio generator
1   power supply: adjustable from approximately 0 to 15 V
2   diodes: 1N4148 or 1N914
4   ½-W resistors: 470 Ω, 1 kΩ, 10 kΩ, 100 kΩ
1   capacitor: 1 μF (10-V rating or better)
1   VOM (analog or digital multimeter)
1   oscilloscope

Procedure

POSITIVE LIMITER

1. In Fig. 9-1, estimate the positive and negative peak output voltages. Record in Table 9-1.
2. Connect the positive limiter of Fig. 9-1. (The 1 kΩ is a dc return in case the source is capacitively coupled.)

Adjust the source to get 1 kHz and 20 V peak-to-peak across the input (equivalent to a peak input of 10 V).

3. Move the oscilloscope leads to output. You should get a positively clipped sine wave. Record the positive and negative peak values in Table 9-1. (You must use the dc input of the oscilloscope.)

NEGATIVE LIMITER

4. In Fig. 9-1, assume the diode polarity is reversed. Record your estimates of the positive and negative output peak voltages in Table 9-1. Reverse the polarity of the diode in your built-up circuit and look at the output waveform. It should be negatively clipped. Record the positive and negative peak values.

COMBINATION LIMITER

5. In Fig. 9-2, estimate the positive and negative peak output voltages. Record your estimates in Table 9-1. Connect the combination limiter.
6. Look at the output waveform. Measure and record the positive and negative peaks.

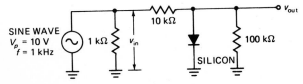

Figure 9-1

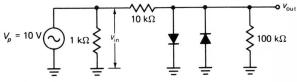

Figure 9-2

BIASED LIMITER

7. In Fig. 9-3, estimate the output peak voltages and record in Table 9-1. Connect the variable limiter of Fig. 9-3.

8. Look at the output with an oscilloscope (dc input). When you vary the dc source, the positive clipping level should vary from a low value to a high value. If it does, write "variable" under positive peak in Table 9-1. Measure and record the negative peak.

PEAK DETECTOR

9. In Fig. 9-4, estimate the dc output voltage, ripple frequency, and peak-to-peak ripple. You may use Eq. (4-10) in the textbook for the latter. Record your estimates in Table 9-2.

10. Connect the peak detector of Fig. 9-4. Adjust the source to get 1 kHz and 10 V peak across the input.

11. Look at the output voltage with the oscilloscope. It should be a dc voltage with an extremely small ripple.

12. Use the VOM to measure the dc output voltage. Record this as V_{dc}.

13. Switch to ac input on the oscilloscope and increase sensitivity until you can measure the ripple accurately. Record the ripple frequency and peak-to-peak ripple.

14. Because the VOM has input resistance on its voltmeter ranges, it may change the resistance across the 1-μF capacitor. While looking at the output ripple on the oscilloscope, connect and disconnect the VOM. What happens to the ripple when the VOM is disconnected? Record "bigger," "same," or "smaller" in Table 9-2.

TROUBLESHOOTING

15. In Fig. 9-2, assume that the left diode is open. Estimate the positive and negative output peak voltages. Record your estimates in Table 9-3.

16. Connect the circuit of Fig. 9-2 with the left diode open. Measure and record the output peak voltages.

17. Repeat Steps 15 and 16 for a shorted diode.

CRITICAL THINKING

18. Assume that the peak voltage is 10 V and the frequency is 5 kHz in Fig. 9-4. Select a filter capacitor (nearest standard size) that produces a peak-to-peak output ripple of approximately 0.5 V. Calculate and record all quantities listed in Table 9-4.

19. Connect the circuit with the new filter capacitor. Adjust the source voltage to 20 V peak-to-peak and the frequency to 5 kHz. Measure and record all quantities listed in Table 9-4.

COMPUTER (OPTIONAL)

20. Repeat Steps 1 to 19 using EWB or an equivalent circuit simulator. Do not record any new values. But make sure that you get reasonable agreement between the EWB measurements and the values recorded earlier.

21. If you are using the CD-ROM version of this book, click on the Assignments menu and select Chap. 4.

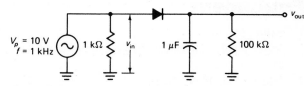

Figure 9-4

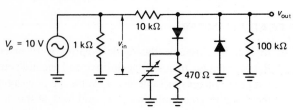

Figure 9-3

Data for Experiment 9

TABLE 9-1. LIMITERS

| | Estimated | | Measured | |
	Pos Peak	Neg Peak	Pos Peak	Neg Peak
Positive limiter				
Negative limiter				
Combination limiter				
Biased limiter				

TABLE 9-2. PEAK DETECTOR

	Estimated	Measured
V_{dc}		
f_{out}		
V_{rip}		
Ripple change	(no entry)	

TABLE 9-3. TROUBLESHOOTING

| | Estimated | | Measured | |
	Pos Peak	Neg Peak	Pos Peak	Neg Peak
Open diode				
Shorted diode				

TABLE 9-4. CRITICAL THINKING

	Calculated	Measured
Capacitance		(no entry)
DC output voltage		
Ripple frequency		
Peak-to-peak ripple		

Questions for Experiment 9

1. In a negative limiter, which of these is the largest?　( )
(a) positive peak;　(b) negative peak;　(c) knee voltage;　(d) crossover voltage.
2. The combination limiter of Fig. 9-2:　( )
(a) puts out a small sine wave;　(b) generates a small squarish wave;　(c) has an adjustable clipping level;　(d) has an output proportional to the input.
3. When the dc source of Fig. 9-3 varies from 0 to 15 V, the positive output peak varies from roughly:　( )
(a) 0 to $V_P/2$;　(b) 0 to V_P;　(c) 0 to $2V_P$;　(d) 0 to 0.7 V.

4. In the combination limiter of Fig. 9-2, which diode approximation is the most reasonable compromise? ( )
   (a) ideal;   **(b)** second;   **(c)** third;   **(d)** fourth.
5. The peak-to-peak ripple out of the peak detector of Fig. 9-4 was approximately ( )
   what percent of the dc output voltage?
   (a) 1%;   **(b)** 5%;   **(c)** 10%;   **(d)** 20%.
6. Briefly explain the operation of the biased combination clipper (Fig. 9-3).

TROUBLESHOOTING

7. Explain why each trouble in Table 9-3 produces the recorded outputs.

8. You are troubleshooting a peak detector like Fig. 9-4. If the output is a half-wave rectified sine wave, what is the trouble?

CRITICAL THINKING

9. Which diode approximation appears to be the best compromise for designing peak detectors? Explain your reasoning.

10. What advantage would a germanium high-frequency signal diode have over a similar silicon diode used as a peak detector? Why is this an advantage?

11. Optional: Instructor's question.

DC Clampers and Peak-to-Peak Detectors

In a dc clamper, a capacitor is charged to approximately the peak input voltage V_P. Depending on the polarity of the charge, the output voltage has a dc component equal to the positive or negative peak input voltage. The output of a positive clamper ideally swings from 0 to $+2V_P$, whereas the output of a negative clamper swings from 0 to $-2V_P$.

A peak-to-peak detector is a cascaded connection of a dc clamper and a peak detector. The dc clamper ideally produces an output that swings from 0 to $2V_P$, and the peak detector produces a dc output of approximately $2V_P$. Since the final dc output equals the peak-to-peak input voltage, the overall circuit is called a peak-to-peak detector.

If a signal source is capacitively coupled, the problem of the dc return may arise with diode and transistor circuits. When the source has to supply more current on one half cycle than the other, its coupling capacitor will charge to approximately the peak of the source voltage. Because of this, you will get unwanted dc clamping of the source signal. To eliminate this unwanted clamping, you can add a dc return. It discharges the coupling capacitor and prevents a dc shift of the output signal.

Required Reading

Chapter 4 (Sec. 4-11) of *Electronic Principles*, 6th ed.

Equipment

1 audio generator
2 diodes: 1N4148 or 1N914
4 ½-W resistors: 1 kΩ, 10 kΩ, 47 kΩ, 100 kΩ
2 capacitors: 1 μF (20-V rating or better)
1 VOM (analog or digital multimeter)
1 oscilloscope

Procedure

POSITIVE CLAMPER

1. In Fig. 10-1, estimate the positive and negative peaks of the output voltage. Record in Table 10-1.

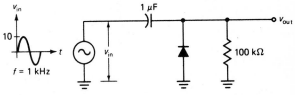

Figure 10-1

2. Connect the positive clamper of Fig. 10-1. Adjust the source to get 1 kHz and 20 V peak-to-peak across the input.
3. With the oscilloscope on dc input, look at the output. It should be a positively clamped sine wave. Measure and record the positive and negative peaks in Table 10-1.
4. Keep the oscilloscope on the output and vary the input voltage. Notice how the negative peak is clamped near zero while the positive peak moves up and down.

NEGATIVE CLAMPER

5. Assume the polarity of the diode in Fig. 10-1 is reversed. Estimate and record the output peaks in Table 10-1.
6. Reverse the polarity of the diode in the built-up circuit. Measure and record the output peaks.

PEAK-TO-PEAK DETECTOR

7. Estimate the dc output voltage and peak-to-peak ripple in Fig. 10-2. You may use Eq. (4-10) for the latter. Record in Table 10-2.
8. Connect the peak-to-peak detector of Fig. 10-2. Adjust the source to get 1 kHz and 20 V peak-to-peak across the input.
9. Look at the voltage across the first diode. It should be a positively clamped signal.
10. Look at the output. It should be a dc voltage with a small ripple. Measure the dc output voltage with a VOM and record in Table 10-2.
11. Switch the oscilloscope to ac input and high sensitivity to measure the ripple. Record V_{rip}.

DC RETURN

12. In Fig. 10-3, the inside of the dashed box simulates a capacitively coupled source. The 1-kΩ resistor is a dc return. Estimate and record the positive peak output voltage (Table 10-3). Visualize the dc return open; estimate and record the positive-peak output voltage.
13. Connect the circuit of Fig. 10-3. Adjust the source to get 1 kHz and 20 V peak-to-peak across the 1-kΩ resistor.

14. Look at the output with the oscilloscope. It should be a half-wave signal. Measure and record the peak value in Table 10-3.
15. Disconnect the 1 kΩ. Measure and record the output peak value.

TROUBLESHOOTING

16. In Fig. 10-2, assume capacitor C_1 is open. Estimate and record the dc output voltage in Table 10-4.
17. Connect the circuit with the foregoing trouble. Measure and record the dc output voltage.
18. Repeat Steps 16 and 17 for each of the remaining troubles listed in Table 10-2.

CRITICAL THINKING

19. The frequency is changed to 2.5 kHz and the load resistor to 47 kΩ in Fig. 10-2. Select a value of output filter capacitance (nearest standard size) that produces a peak-to-peak ripple of approximately 0.1 V. Calculate all quantities listed in Table 10-5.

COMPUTER (OPTIONAL)

20. Repeat Steps 1 to 19 using EWB or an equivalent circuit simulator. Do not record any new values. But make sure that you get reasonable agreement between the EWB measurements and the values recorded earlier.
21. If you are using the CD-ROM version of this book, click on the Assignments menu and select Chap. 4.

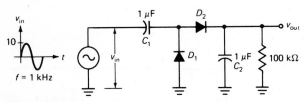

Figure 10-2

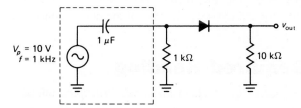

Figure 10-3

Data for Experiment 10

TABLE 10-1. CLAMPERS

	Estimated		Measured	
	Pos Peak	Neg Peak	Pos Peak	Neg Peak
Positive clamper				
Negative clamper				

TABLE 10-2. PEAK-TO-PEAK DETECTOR

	Estimated	Measured
V_{dc}		
V_{rip}		

TABLE 10-3. DC RETURN

	Estimated V_P	Measured V_P
With dc return		
Without dc return		

TABLE 10-4. TROUBLESHOOTING

	Estimated V_{dc}	Measured V_{dc}
Open C_1		
Short C_1		
Open C_2		
Short C_2		
Open D_1		
Short D_1		
Open D_2		
Short D_2		

TABLE 10-5. CRITICAL THINKING

	Calculated	Measured
Capacitance		
DC output voltage		
Ripple frequency		
Peak-to-peak ripple		

Questions for Experiment 10

1. If the diode of Fig. 10-1 is reversed, the output will be: ( )
   (a) positively clamped;   **(b)** negative clamped;   **(c)** half-wave rectified;
   (d) peak rectified.

2. If V_P is 10 V in Fig. 10-2, the maximum positive voltage across the first diode is   ( )
   approximately:
   (a) 5 V;   **(b)** 10 V;   **(c)** 15 V;   **(d)** 20 V.

3. If V_P is 10 V in Fig. 10-2, the dc output voltage is ideally: ( )
   (a) 0 V;   **(b)** 5 V;   **(c)** 10 V;   **(d)** 20 V.

4. In Fig. 10-2, the peak-to-peak ripple is approximately what percent of dc output   ( )
   voltage?
   (a) 0;   **(b)** 1%;   **(c)** 5%;   **(d)** 63.6%.

5. When the dc return of Fig. 10-3 is disconnected, which of the following is true?   ( )
   (a) The capacitor charges to approximately $2V_P$;   **(b)** current flows easily in
   the reverse diode direction;   **(c)** the diode conducts briefly near each positive
   peak;   **(d)** the diode eventually stops conducting.

6. Explain how a positive dc clamper works.

TROUBLESHOOTING

7. What dc output voltage did you get when C_2 was opened in Fig. 10-2? Explain why this
   happened.

8. A group of technicians is gathered around a circuit that works with one signal generator
   but not with another. No one can figure out why one generator works but not the other. Explain what is probably happening.

9. Optional: Instructor's question.

10. Optional: Instructor's question.

Voltage Doublers

A voltage multiplier produces a dc voltage equal to a multiple of the peak input voltage. Voltage multipliers are useful with high-voltage/low-current loads. With a voltage doubler, you get twice as much dc output voltage as you do from a standard peak rectifier. This is useful when you are trying to produce high voltages (several hundred volts or more) because higher secondary voltages result in bulkier transformers. At some point, a designer may prefer to use voltage doublers instead of bigger transformers. With a voltage tripler, the dc voltage is approximately three times the peak input voltage. As the multiple increases, the peak-to-peak ripple gets worse.

In this experiment, you will build half-wave and full-wave voltage doublers. You will measure the dc output voltage and peak-to-peak ripple of these circuits to verify the operation described in the textbook.

Required Reading

Chapter 4 (Sec. 4-12) of *Electronic Principles,* 6th ed.

Equipment

1 transformer: 12.6 V ac center-tapped (Triad F-25X or equivalent) with fused line cord
2 silicon diodes: 1N4001 (or equivalent)
1 ½-W resistor: 1 kΩ
2 capacitors: 470 μF (25-V rating or better)
1 VOM (analog or digital multimeter)
1 oscilloscope

Procedure

HALF-WAVE DOUBLER

1. Measure the resistance of the primary and secondary windings. Record in Table 11-1.
2. In Fig. 11-1, assume that the rms secondary voltage is 12.6 V. Calculate and record the quantities listed in Table 11-2. Use Eq. (4-8) in the textbook to calculate the peak-to-peak ripple.
3. Connect the circuit.
4. Measure and record all the quantities listed in Table 11-2.

FULL-WAVE DOUBLER

5. Repeat Steps 2 through 4 for the full-wave doubler of Fig. 11-2. Use Table 11-3 to record your data. When calculating the peak-to-peak ripple, notice that the load resistor is in parallel with two capacitors in series.

TROUBLESHOOTING

6. Assume that capacitor C_1 is open in Fig. 11-1.
7. Estimate the dc load voltage, ripple frequency, and peak-to-peak ripple. Record your rough estimates in Table 11-4.
8. Connect the circuit with the foregoing trouble. Measure and record the quantities of Table 11-4.
9. Assume that diode D_2 is open in Fig. 11-1. Repeat Steps 7 and 8.

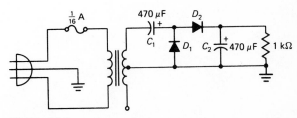

Figure 11-1

CRITICAL THINKING

10. Select a filter capacitor (nearest standard size) for the circuit of Fig. 11-1 to get a peak-to-peak ripple of approximately 10 percent of load voltage for an R_L of 3.9 kΩ. Calculate and record the quantities of Table 11-5. Record your design value for capacitance here:

$$C =$$

11. Connect your circuit. Measure and record the quantities of Table 11-5.

COMPUTER (OPTIONAL)

12. Repeat Steps 1 to 11 using EWB or an equivalent circuit simulator. Do not record any new values. But make sure that you get reasonable agreement between the EWB measurements and the values recorded earlier.

13. If you are using the CD-ROM version of this book, click on the Assignments menu and select Chap. 4.

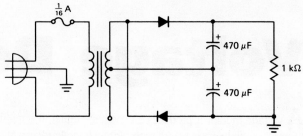

Figure 11-2

Data for Experiment 11

TABLE 11-1. TRANSFORMER RESISTANCES

$R_{pri} =$ _____

$R_{sec} =$ _____

TABLE 11-2. HALF-WAVE DOUBLER

	Calculated	Measured
Half rms secondary voltage		
DC output voltage		
Ripple frequency		
Peak-to-peak ripple		

TABLE 11-3. FULL-WAVE DOUBLER

	Calculated	Measured
Half rms secondary voltage		
DC output voltage		
Ripple frequency		
Peak-to-peak ripple		

TABLE 11-4. TROUBLESHOOTING

	Estimated				Measured		
	V_{dc}	f_{out}	V_{rip}		V_{dc}	f_{out}	V_{rip}
Open C_1							
Open D_2							
Open C_2							

TABLE 11-5. CRITICAL THINKING

	Calculated	Measured
Half rms secondary voltage		
DC load voltage		
Ripple frequency		
Peak-to-peak ripple		

Questions for Experiment 11

1. In this experiment the dc output voltage from the half-wave doubler was approximately equal to: ( )
   (a) peak primary voltage;   (b) rms secondary voltage;   (c) double the peak secondary voltage;   (d) double the peak voltage driving the half-wave doubler.

2. The ripple frequency of a half-wave doubler was: ( )
   (a) 60 Hz;   (b) 120 Hz;   (c) 240 Hz;   (d) 480 Hz.

3. The full-wave doubler has a ripple frequency of: ( )
   (a) 60 Hz;   (b) 120 Hz;   (c) 240 Hz;   (d) 480 Hz.

4. The peak-to-peak ripple of a full-wave doubler compared with a half-wave doubler is: ( )
   (a) half;   (b) the same;   (c) twice as much.

5. In Fig. 11-1, D_1 and C_1 act like a: ( )
   (a) positive clamper;   (b) negative clamper;   (c) peak detector; (d) diode clamp.

6. Briefly explain how the full-wave doubler of Fig. 11-2 works.

TROUBLESHOOTING

7. Explain why the peak-to-peak ripple is so large for an open C_2 in Table 11-4.

8. Suppose either filter capacitor in Fig. 11-2 is shorted. Explain what happens to the nearest diode.

CRITICAL THINKING

9. Justify your design; that is, why did you use the filter capacitor you selected?

10. Assume that the primary resistance is 30 Ω and the secondary resistance is 1 Ω in Fig. 11-2. The primary voltage is 115 V, and the secondary voltage is 12.6 V. What is the Thevenin resistance facing either filter capacitor? Ignore the bulk resistance of the diodes in this calculation.

The Zener Diode

Ideally, a zener diode is equivalent to a dc source when operating in the breakdown region. To a second approximation, it is like a dc source with a small internal impedance. Its main advantage is the approximately constant voltage appearing across it. In this experiment you will get data for the zener voltage and zener resistance.

Required Reading

Chapter 5 (Secs. 5-1 to 5-7) of *Electronic Principles*, 6th ed.

Equipment

1 power supply: adjustable from approximately 0 to 15 V
1 zener diode: 1N753
1 ½-W resistor: 180 Ω
1 VOM (analog or digital multimeter)

Procedure

ZENER VOLTAGE

1. Measure the diode's forward and reverse resistance on one of the middle resistance ranges. The reverse/forward resistance ratio should be at least 1000 : 1.
2. The 1N753 has a nominal zener voltage of 6.2 V. In Fig. 12-1, estimate and record the output voltage for each input voltage listed in Table 12-1.
3. Connect the circuit of Fig. 12-1. Measure and record the output voltage for each input voltage of Table 12-1.

ZENER RESISTANCE

4. With the data of Table 12-1, calculate and record the zener current in Fig. 12-1 for each entry of Table 12-2.
5. With Eq. (5-17) in the textbook, calculate the zener resistance for V_{in} = 10 V. (Use the voltage and current changes between 8 and 12 V.)
6. Calculate and record the zener resistance for V_{in} = 12 V.

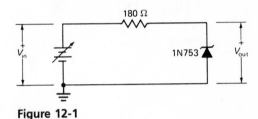

Figure 12-1

CURVE TRACER

7. If a curve tracer is available, display the forward and reverse zener curves.

TROUBLESHOOTING

8. In Fig. 12-1, assume that V_{in} is 9 V and estimate the output voltage for a shorted zener diode. Record your answer in Table 12-3.
9. Estimate and record the output voltage for an open zener diode.
10. Estimate and record V_{out} for an open resistor.
11. Assume the polarity of the zener diode is reversed. Estimate and record the output voltage for this trouble.
12. Connect the circuit with each of the foregoing troubles. Measure and record V_{out} for a V_{in} of 9 V.

CRITICAL THINKING

13. Select a current-limiting resistor to produce a zener current of approximately 16.5 mA when V_{in} is 14 V. Record your design value at the top of Table 12-4. Connect the circuit with your design value of R_S. Measure and record the output voltage for each input voltage listed in Table 12-4.

14. Calculate and record the zener current for each input voltage of Table 12-4. Calculate and record the zener resistance for an input voltage of 12 V.

APPLICATION (OPTIONAL)

15. The Radio Shack buzzer RS273-026 has an input voltage range of 6 to 16 V with a rated current of 10 mA at 12 V. Given a 20-V power supply, we can use a 1N757 to reduce the voltage applied to this buzzer. As a demonstration, connect the circuit of Fig. 12-2. You

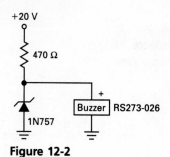

Figure 12-2

should measure approximately 9 V across the zener diode, and you should hear a buzz.

COMPUTER (OPTIONAL)

16. Repeat Steps 1 to 14 using EWB or an equivalent circuit simulator. Do not record any new values. But make sure that you get reasonable agreement between the EWB measurements and the values recorded earlier.

17. If you are using the CD-ROM version of this book, click on the Assignments menu and select Chap. 5.

ADDITIONAL WORK (OPTIONAL)

18. Your instructor will give you a diode with a zener voltage between 2.4 and 12 V. You may assume that the diode can safely handle up to 20 mA of current. Figure out how to find the zener voltage for any current less than 20 mA. After you have measured the zener voltage, figure out how to vary the zener current from 1 to 10 mA. Finally, take enough measurements to graph I_Z versus V_Z over the foregoing current range.

Data for Experiment 12

TABLE 12-1. DATA FOR ZENER DIODE

V_{in}	Estimated V_{out}	Measured V_{out}
0		
2 V		
4 V		
6 V		
8 V		
10 V		
12 V		
14 V		

TABLE 12-2. ZENER RESISTANCE

V_{in}	Calculated I_Z	Calculated R_Z
0		(no entry)
2 V		(no entry)
4 V		(no entry)
6 V		(no entry)
8 V		(no entry)
10 V		
12 V		
14 V		(no entry)

TABLE 12-3. TROUBLESHOOTING

Trouble	Estimated V_{dc}	Measured V_{dc}
Shorted diode		
Open diode		
Open resistor		
Reversed diode		

TABLE 12-4. CRITICAL THINKING: R_S = _____

V_{in}	Measured V_{out}	Calculated I_Z	Calculated R_Z
10 V			(no entry)
12 V			
14 V			(no entry)

Questions for Experiment 12

1. In Fig. 12-1, the zener current and the current through the 180-Ω resistor are: ( )
   (a) equal;   (b) almost equal;   (c) much different.
2. The zener diode starts to break down when the input voltage is approximately: ( )
   (a) 4 V;   (b) 6 V;   (c) 8 V;   (d) 10 V.
3. When V_{in} is less than 6 V, the output voltage is: ( )
   (a) approximately constant;   (b) negative;   (c) the same as the input.
4. When V_{in} is greater than 8 V, the output voltage is: ( )
   (a) approximately constant;   (b) negative;   (c) the same as the input.
5. The calculated zener resistances were closest to: ( )
   (a) 1 Ω;   (b) 2 Ω;   (c) 7 Ω;   (d) 20 Ω.
6. Explain why the zener diode is called a constant-voltage device.

TROUBLESHOOTING

7. Explain the value of output voltage you got when the zener diode was open.

8. Explain the value of output voltage you got when the zener diode was reversed.

CRITICAL THINKING

9. How and why did you select the value of current-limiting resistance?

10. Optional. Instructor's question.

50

The Zener Regulator

In a zener voltage regulator a load resistor is in parallel with a zener diode. As long as the zener diode operates in the breakdown region, the load voltage is approximately constant and equal to the zener voltage. In a stiff zener regulator, the zener resistance is less than 1/100 of the series resistance and less than 1/100 of the load resistance. By meeting the first condition, a zener regulator attenuates the input ripple by a factor of at least 100. By meeting the second condition, a zener regulator appears like a stiff voltage source to the load resistance.

In this experiment you will build a split supply with regulated positive and negative output voltages. This will allow you to verify the operation of a zener regulator as described in your textbook.

Required Reading

Chapter 5 (Secs. 5-1 to 5-7) of *Electronic Principles*, 6th ed.

Equipment

1   center-tapped transformer, 12.6 V ac (Triad F-25X or equivalent) with fused line cord
4   silicon diodes: 1N4001 (or equivalent)
2   zener diodes: 1N753
8   ½-W resistors: two 150 Ω, two 470 Ω, two 4.7 kΩ, two 47 kΩ
2   capacitors: 470 μF (25-V rating or better)

1   VOM (analog or digital multimeter)
1   oscilloscope

Procedure

SPLIT SUPPLY

1. A 1N753 has a nominal zener voltage of 6.2 V. In Fig. 13-1, calculate the input and output voltages for each zener regulator. (The input voltages are across the filter capacitors.) Record your answers in Table 13-1.
2. Connect the split supply of Fig. 13-1.
3. Measure the input and output voltages of each zener regulator. Record your data in Table 13-1.

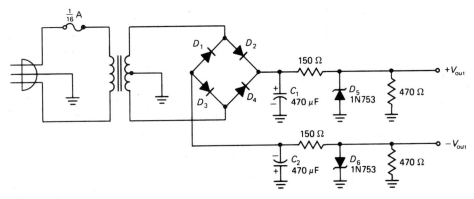

Figure 13-1

VOLTAGE REGULATION

4. Estimate and record the output voltages in Fig. 13-1 for each of the load resistors listed in Table 13-2.
5. Connect the circuit. Measure and record the output voltages for the load resistances of Table 13-2.

RIPPLE ATTENUATION

6. For each load resistance listed in Table 13-3, calculate and record the peak-to-peak ripple across the upper filter capacitor of Fig. 13-1. Also calculate and record the peak-to-peak ripple at the positive output. (Assume a zener resistance of 7 Ω.)
7. For each load resistance of Table 13-3, measure and record the peak-to-peak ripple at the input and output of the positive zener regulator.

TROUBLESHOOTING

8. Assume that the center tap of Fig. 13-1 is open.
9. Estimate the output voltages for the foregoing trouble. Record your answers in Table 13-4.
10. Connect the circuit with the foregoing trouble. Measure and record the output voltages. Remove the trouble.
11. Repeat Steps 9 and 10 for the other troubles listed in Table 13-4.

CRITICAL THINKING

12. Design a two-stage voltage regulator similar to Fig. 5-7 of your textbook to meet these specifications: pre-regulator output is a nominal +12.4 V, final output is a nominal +6.2 V, current in preregulator series resistor is 40 mA, current in final series resistor is 20 mA, and ripple attenuation is at least 300. Assume a zener resistance of 7 Ω for each diode. Use the 1N753s and any additional resistors as required. Draw your final design at the bottom of Table 13-5.
13. Calculate and record the dc voltage and peak-to-peak ripple at the preregulator input, regulator input, and final output (Table 13-5).
14. Check with the instructor about the safety of your design. Then connect your design with a load resistance of 470 Ω. Measure all dc voltages and ripples listed in Table 13-5. Record your data.

COMPUTER (OPTIONAL)

15. Repeat Steps 1 to 14 using EWB or an equivalent circuit simulator. Do not record any new values. But make sure that you get reasonable agreement between the EWB measurements and the values recorded earlier.
16. If you are using the CD-ROM version of this book, click on the Assignments menu and select Chap. 5.

ADDITIONAL WORK (OPTIONAL)

17. Have another student insert one of the following troubles into the circuit: *open* any diode, resistor, capacitor, fuse, or connecting wire. Use only voltage readings of a DMM or oscilloscope to troubleshoot.
18. Repeat Step 17 several times until you are confident you can troubleshoot the circuit for open components.

Data for Experiment 13

TABLE 13-1. SPLIT SUPPLY

	Calculated		Measured	
	V_{in}	V_{out}	V_{in}	V_{out}
Positive regulator				
Negative regulator				

TABLE 13-2. VOLTAGE REGULATION

R_L	Estimated		Measured	
	$+V_{out}$	$-V_{out}$	$+V_{out}$	$-V_{out}$
470 Ω				
4.7 kΩ				
47 kΩ				

TABLE 13-3. RIPPLE

R_L	Calculated V_{rip}		Measured V_{rip}	
	In	Out	In	Out
470 Ω				
47 kΩ				

TABLE 13-4. TROUBLESHOOTING

	Estimated		Measured	
	$+V_{out}$	$-V_{out}$	$+V_{out}$	$-V_{out}$
Open CT				
Open D_1				
Open D_6				

TABLE 13-5. CRITICAL THINKING

	Calculated		Measured	
	V_{dc}	V_{rip}	V_{dc}	V_{rip}
Preregulator input				
Regulator input				
Regulator output				

Draw your design here:

Questions for Experiment 13

1. A split supply has: ( )
   (a) only one output voltage;   (b) only a positive output voltage;   (c) only a negative output voltage;   (d) positive and negative outputs.

2. The value of V_{in} to the positive zener regulator is closest to: ( )
   (a) 5 V;   (b) 10 V;   (c) 15 V;   (d) 20 V.

3. When R_L increases in Table 13-2, the measured positive output voltage: ( )
   (a) decreases slightly;   (b) remains the same;   (c) increases slightly.

4. Theoretically, the positive zener regulator of Fig. 13-1 attenuates the ripple by a   ( )
   factor of approximately:
   (a) 10;   (b) 20;   (c) 50;   (d) 100.

5. The current through either series resistor of Fig. 13-1 is closest to: ( )
   (a) 5 mA;   (b) 10 mA;   (c) 15 mA;   (d) 20 mA.

6. Explain how the positive zener regulator of Fig. 13-1 works.

TROUBLESHOOTING

7. Explain why the circuit of Fig. 13-1 continued to work even though you opened the center tap.

8. Explain why the circuit of Fig. 13-1 still works with an open D_2.

CRITICAL THINKING

9. Explain why the measured ripples did not agree exactly with the calculated ripples in your design.

10. Optional. Instructor's question.

Optoelectronic Devices

In a forward-biased LED, heat and light are radiated when free electrons and holes recombine at the junction. Because the LED material is semitransparent, some of the light escapes to the surroundings. LEDs have a typical voltage drop from 1.5 and 2.5 V for currents between 10 and 50 mA. The exact voltage drop depends on the color, tolerance, and other factors. For troubleshooting and design, we will use the second diode approximation with a knee of 2 V.

An LED array is a group of LEDs that display numbers, letters, or other symbols. The most common LED array is the seven-segment display. It contains seven rectangular LEDs. Each LED is called a segment because it forms one part of the character being displayed. By activating one or more LEDs, we can form any digit from 0 through 9.

An optocoupler combines an LED and a photodetector in a single package. The light from the LED hits the photodetector. This produces an output voltage that depends on the amount of current through the LED. If the LED current has an ac variation, V_{out} will have an ac variation. The key advantage of an optocoupler is the electrical isolation between the LED circuit and the output circuit, typically in thousands of megohms.

Required Reading

Chapter 5 (Sec. 5-3) of *Electronic Principles*, 6th ed.

Equipment

2 power supplies: one at 15 V, another adjustable from approximately 0 to 15 V
2 LEDs: L53RD and L53GD (or equivalent red and green LEDs)
2 1-W resistors: 270 Ω
Seven-segment display: TIL312 (or equivalent)
Optocoupler: 4N26 (or equivalent)
1 VOM (analog or digital multimeter)

Procedure

DATA FOR A RED LED

1. Examine the red LED. Notice that one side of the package has a flat edge. This indicates the cathode side. (With many LEDs, the cathode lead is slightly shorter than the anode lead. This shorter lead is another way to identify the cathode.)

2. Connect the circuit of Fig. 14-1 using a red LED.

3. Adjust source voltage V_S to get 10 mA through the LED. Record the corresponding LED voltage in Table 14-1.

4. Adjust the source voltage and set up the remaining currents listed in Table 14-1. Record each LED voltage.

DATA FOR A GREEN LED

5. Replace the red LED by a green LED in the circuit of Fig. 14-1.

6. Repeat Steps 3 and 4 for the green LED.

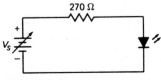

Figure 14-1

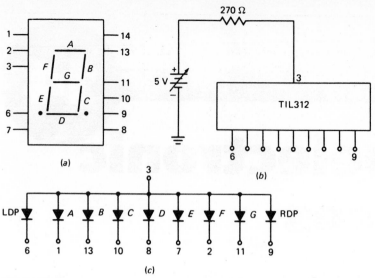

Figure 14-2

USING A SEVEN-SEGMENT DISPLAY

7. Figure 14-2a shows the pinout for the seven-segment display used in this experiment (top view). It includes a left decimal point (LDP) and a right decimal point (RDP). Connect the circuit of Fig. 14-2b.
8. Figure 14-2c shows the schematic diagram for a TIL312. Ground pins 1, 10, and 13. If the circuit is working correctly, digit 7 will be displayed.
9. Disconnect the grounds on pins 1, 10, and 13.
10. Refer to Fig. 14-2a and c. Which pins should you ground to display a zero? Ground these pins and if the circuit is working correctly, enter the pin numbers in Table 14-2.
11. Repeat Step 10 for the remaining digits, 1 through 9, and the decimal points.

THE TRANSFER GRAPH OF AN OPTOCOUPLER

12. Connect the circuit of Fig. 14-3. Adjust the source voltage to 2 V. Measure and record the output voltage (Table 14-3).
13. Repeat Step 12 for the source voltages shown in Table 14-3. Record the corresponding output voltages.

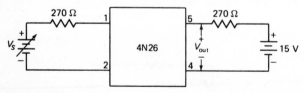

Figure 14-3

TROUBLESHOOTING

14. If V_S is 15 V in Fig. 14-1, estimate the voltage across the red LED if it is open. Record your answer in Table 14-4. Similarly, estimate the LED voltage if the LED is shorted.
15. Connect the circuit with a source voltage of 15 V and a red LED. Measure and record the LED voltage for the troubles listed in Table 14-4.

CRITICAL THINKING

16. Select a current-limiting resistor for a red LED in Fig. 14-1 that sets up a current of approximately 20 mA when the source voltage is 15 V. Record your design value in Table 14-5. Calculate and record the LED current and voltage.
17. Connect your design. Measure and record the LED current and voltage.
18. Repeat Steps 16 and 17 for the green LED.

APPLICATION (OPTIONAL)

19. Measure the resistance of a few cadmium-sulfide photocells such as the Radio Shack 276-1657, an assorted package of photoresistors, devices whose resistance changes with the amount of incoming light. Notice how the resistance decreases when you expose the photocell to light and how it increases when you cover the photocell.
20. Build the circuit of Fig. 14-4. The LED should be lit. Cover the photocell and notice how the LED either goes out or becomes less bright. Repeat this demonstration with a few other photocells. *Note:* Make sure

56

that the photocell is pointing toward a light source. If the LED does not light up, the photocell resistance is too high. Either try using another photocell or slowly increase the supply voltage from 10 V toward a maximum of 30 V until the LED lights up enough to be seen. Do not use too much voltage because the LED may be destroyed if the current exceeds 50 mA. If you like, use an ammeter in series with the LED and set the current to approximately 10 mA.

COMPUTER (OPTIONAL)

21. Repeat Steps 1 to 6 using EWB or an equivalent circuit simulator. Do not record any new values. But make sure that you get reasonable agreement between the EWB measurements and the values recorded earlier.

22. If you are using the CD-ROM version of this book, click on the Assignments menu and select Chap. 5.

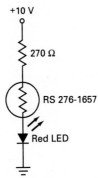

Figure 14-4

Data for Experiment 14

TABLE 14-1. LED DATA

I	V_{red}	V_{green}
10 mA		
20 mA		
30 mA		
40 mA		

TABLE 14-2. SEVEN-SEGMENT INDICATOR

Display	Pins grounded
0	
1	
2	
3	
4	
5	
6	
7	
8	
9	
LDP	
RDP	

TABLE 14-3. OPTOCOUPLER

V_S	V_{out}
2 V	
4 V	
6 V	
8 V	
10 V	
12 V	
14 V	

TABLE 14-4. TROUBLESHOOTING

	Estimated V_{LED}	Measured V_{LED}
Open LED		
Shorted LED		

TABLE 14-5. CRITICAL THINKING

		Calculated		Measured	
	R	I_{LED}	V_{LED}	I_{LED}	V_{LED}
Red LED					
Green LED					

Questions for Experiment 14

1. The voltage drop across the red LED for a current of 30 mA was closest to: ( )
   (a) 0 V;   **(b)** 1 V;   **(c)** 2 V;   **(d)** 4 V.
2. The voltage drop across the green LED for a current of 30 mA was closest to: ( )
   (a) 0 V;   **(b)** 1 V;   **(c)** 2 V;   **(d)** 4 V.
3. To display a 1 on the seven-segment indicator, which pins did you ground? ( )
   (a) 1;   **(b)** 1 and 10;   **(c)** 10 and 13;   **(d)** 2, 7, and 8.
4. In Fig. 14-2, which of the following is true? ( )
   (a) LED brightness decreases as more segments are lit;   **(b)** all segments are equally bright at all times;   **(c)** number 8 was brighter than number 1.
5. When the source voltage increases in Fig. 14-3, the output voltage: ( )
   (a) decreases;   **(b)** stays the same;   **(c)** increases.
6. Explain how the LED array of Fig. 14-2 works. Include a discussion of the LED current versus the total current.

TROUBLESHOOTING

7. Why was the LED voltage large when the LED was open?

CRITICAL THINKING

8. Explain how you calculated the current-limiting resistor for the red LED in Table 14-5.

9. It is possible to get equal brightness of all numbers with the LED array of Fig. 14-2. How can this be done?

10. Optional. Instructor's question.

60

The CE Connection

As an approximation of transistor behavior, we use the Ebers-Moll model: the emitter diode acts like a rectifier diode, while the collector diode acts like a controlled-current source. The voltage across the emitter diode of a small-signal transistor is typically 0.6 to 0.7 V. For most troubleshooting and design, we will use 0.7 V for the V_{BE} drop. In this experiment, you will get data for calculating the α_{dc}, β_{dc}, and the V_{BE} drop.

When the maximum ratings of a transistor are exceeded, it can be damaged in several ways. The most common transistor trouble is a collector-emitter short where both the emitter diode and the collector diode are shorted. Another common transistor trouble is the collector-emitter open where both the emitter diode and the collector diode are open. Besides the foregoing, it is possible to have only one diode shorted, only one diode open, a leaky diode, etc.

To keep the troubleshooting straightforward, we will emphasize the two most common troubles: the collector-emitter short and the collector-emitter open. We will simulate a collector-emitter short by putting a jumper between the collector, base, and emitter; this shorts all three terminals together. We will simulate the collector-emitter open by removing the transistor from the circuit; this opens both diodes.

Required Reading

Chapter 6 (Secs. 6-1 to 6-9) of *Electronic Principles*, 6th ed.

Equipment

1 power supply: 15 V
3 ½-W resistors: 100 Ω, 1 kΩ, 470 kΩ
3 transistors: 2N3904 (or almost any small-signal *npn* silicon transistor)
1 VOM (analog or digital multimeter)

Procedure

OHMMETER TESTS

1. Measure the resistance between the collector and emitter of one of the transistors. This resistance should be extremely high (hundreds of megohms) in either direction.
2. Measure the forward and reverse resistance of the base-emitter diode and the collector-base diode. For

both diodes, the reverse/forward resistance ratio should be at least 1000:1.
3. Repeat Steps 1 and 2 for two other transistors.

TRANSISTOR CHARACTERISTICS

4. Connect the circuit of Fig. 15-1, using one of the transistors.
5. Measure and record V_{BE} and V_{CE} in Table 15-1.
6. Measure and record I_C and I_B in Table 15-1.
7. Calculate the values of V_{CB}, I_E, α_{dc}, and β_{dc} in Fig. 15-1. Record in Table 15-2.
8. Repeat Steps 4 to 7, using a second transistor.
9. Repeat Steps 4 to 7, using a third transistor.

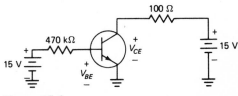

Figure 15-1

10. If a curve tracer is available, display the collector curves of all three transistors. Notice the differences in β_{dc}, breakdown voltages, etc.

TROUBLESHOOTING

11. In Fig. 15-2, estimate and record the collector-to-ground voltage V_C for each trouble listed in Table 15-3. Note: To simulate a collector-emitter short, put a jumper between the collector, emitter, and base so that all three terminals are shorted together. To simulate a collector-emitter open, remove the transistor from the circuit.
12. Connect the circuit with each of the foregoing troubles. Measure and record the collector voltage for each trouble.

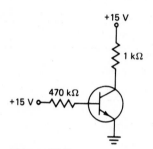

Figure 15-2

CRITICAL THINKING

13. Connect the circuit of Fig. 15-2 and measure the collector-to-ground voltage.
14. With the data of Step 13, calculate β_{dc} and select a base resistance that will produce a collector voltage of approximately half the supply voltage. Record the nearest standard resistance and the calculated collector voltage in Table 15-4.
15. Connect the circuit with your design value of base resistance. Complete the entries of Table 15-4.

COMPUTER (OPTIONAL)

16. Repeat Steps 1 to 15 using EWB or an equivalent circuit simulator. Do not record any new values. But make sure that you get reasonable agreement between the EWB measurements and the values recorded earlier.
17. If you are using the CD-ROM version of this book, click on the Assignments menu and select Chap. 6.

ADDITIONAL WORK (OPTIONAL)

18. Repeat Steps 4 through 7 for a base resistance of 220 kΩ. Record all values on a separate piece of paper and compare these values to those recorded earlier.
19. Repeat Steps 4 through 7 for a base resistance of 100 kΩ. Record all values on a separate piece of paper and compare these values to those recorded earlier. What did you learn?

Data for Experiment 15

TABLE 15-1. TRANSISTOR VOLTAGES AND CURRENTS

Transistor	V_{BE}	V_{CE}	I_B	I_C
1				
2				
3				

TABLE 15-2. CALCULATIONS

Transistor	V_{CB}	I_E	α_{dc}	β_{dc}
1				
2				
3				

TABLE 15-3. TROUBLESHOOTING

Trouble	Estimated V_C	Measured V_C
Open 470 kΩ		
Shorted 1 kΩ		
Open 1 kΩ		
Shorted collector-emitter		
Open collector-emitter		

TABLE 15-4. CRITICAL THINKING

	R_B	V_C
Calculated		
Measured		

Questions for Experiment 15

1. The V_{BE} drop of the transistors was closest to: ( )
  (a) 0 V;   **(b)** 0.3 V;   **(c)** 0.7 V;   **(d)** 1 V.

2. The α_{dc} of all transistors was very close to: ( )
  (a) 0;   **(b)** 1;   **(c)** 5;   **(d)** 20.

3. The β_{dc} of all transistors was greater than: ( )
  (a) 0;   **(b)** 1;   **(c)** 5;   **(d)** 20.

4. This experiment proves that collector current is much greater than: ( )
  (a) collector voltage;   **(b)** emitter current;   **(c)** base current;   **(d)** 0.7 V.

5. The transistors were silicon because: ( )
  (a) V_{BE} was approximately 0.7 V;   **(b)** I_C is much greater than I_B;   **(c)** the collector diode was reverse-biased;   **(d)** β_{dc} was much greater than unity.

6. What did you learn about the V_{BE} drop and the relation of collector current to base current?

TROUBLESHOOTING

7. What collector voltage did you measure when the base resistor was open? Explain why this voltage existed at the collector.

8. Briefly explain why the collector voltage is approximately zero when a transistor has a collector-emitter short.

CRITICAL THINKING

9. Explain how you calculated the base resistance in Table 15-4.

10. Why is it that increasing or decreasing V_{CC} or R_C in Fig. 15-1 does not change the collector current?

16

Transistor Operating Regions

To determine the region of transistor operation (cutoff, saturation, or active), ask yourself two questions: First, is there base current? Second, is there collector current? To have base current, two conditions are necessary: A complete path must exist for base current, and a voltage must be applied somewhere in this path. Similar conditions apply to the collector circuit.

When there is no base current, the transistor goes into cutoff. For instance, if the base resistor of Fig. 16-1 were open, the path for base current would be broken. In this case, base current would be zero. Since the collector current equals the dc beta times the base current, the collector current would be ideally zero and the transistor would be cut off. In a circuit like Fig. 16-1, no collector current implies that $V_{CE} = V_{CC} = 10$ V.

When there is base current, V_{BE} is approximately 0.7 V. In this case, the transistor may operate in any of the three regions: active, saturation, or cutoff. As a guide, a small-signal transistor is operating in the active region when the collector-emitter voltage V_{CE} is greater than 1 V but less than V_{CC}. Saturation occurs somewhere below 1 V, depending on the transistor type. For a typical small-signal transistor, $V_{CE(sat)}$ is 0.1 to 0.2 V.

In this experiment, you will measure voltages for different transistor circuits. Then, you will examine the data and determine whether the transistor operates in the active, saturation, or cutoff region.

Required Reading

Chapter 6 of *Electronic Principles*, 6th ed.

Equipment

1 power supply: adjustable to 10 V
1 VOM (analog or digital multimeter)
1 transistor: 2N3904
3 ½-W resistors: 1 kΩ, 10 kΩ, 470 kΩ

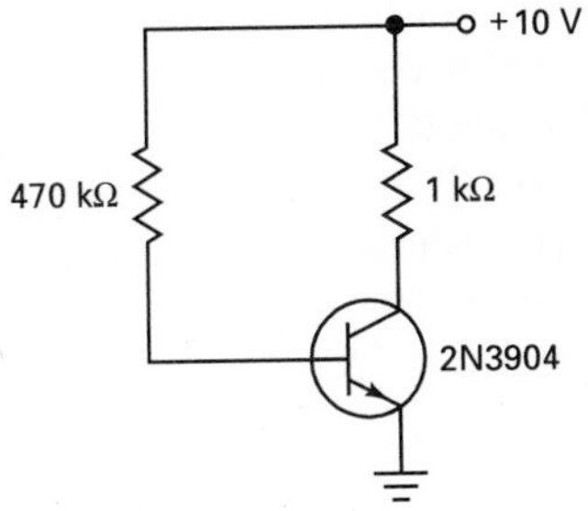

Figure 16-1

Procedure

1. Assume a dc beta of 200 in Fig. 16-1. Calculate V_{BE} and V_{CE}. In which of the three regions is the transistor operating? Record the voltages and the region in Table 16-1.
2. Build the circuit of Fig. 16-1. Measure V_{BE} and V_{CE}. In which of the three regions is the transistor oper-

ating? Record the voltages and the region in Table 16-1.
3. Repeat Steps 1 and 2 for Fig. 16-2. Use Table 16-2.
4. Repeat Steps 1 and 2 for Fig. 16-3. Use Table 16-3.
5. Repeat Steps 1 and 2 for Fig. 16-4. Use Table 16-4.
6. Figure 16-5 shows a circuit with four transistors, labeled Q_1 to Q_4. Estimate the values of V_{BE} and V_{CE} for each transistor, and then determine the operating

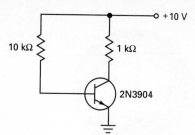

Figure 16-2

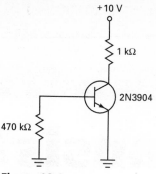

Figure 16-4

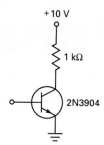

Figure 16-3

region for the transistor. Record your voltage estimates and the operating regions in Table 16-5.

TROUBLESHOOTING

7. Suppose the base resistor is open in Fig. 16-1. What are the values of V_{BE} and V_{CE}? In what region is the transistor operating? Record your answers in Table 16-6 next to "R_{BO}."

8. In attempting to connect the circuit of Fig. 16-1, somebody grounds the base resistor instead of connecting it to the supply voltage. This results in the circuit of Fig. 16-4. What are the values of V_{BE} and V_{CE}? In what region is the transistor operating? Record your answers in Table 16-6 next to "R_{BG}."

9. Suppose the collector resistor is open in Fig. 16-1. What are the values of V_{BE} and V_{CE}? In what region is the transistor operating? Record your answers in Table 16-6 next to "R_{CO}."

10. Suppose the collector resistor is shorted in Fig. 16-1. What are the values of V_{BE} and V_{CE}? In what region is the transistor operating? Record your answers in Table 16-6 next to "R_{CS}."

11. Build the circuit of Fig. 16-1 with a shorted collector resistance. Then, measure V_{BE} and V_{CE} to verify reasonable agreement with the values recorded for R_{CS} in Table 16-6.

12. Build the circuit of Fig. 16-1 with an open collector resistance. Then, measure V_{BE} and V_{CE} to verify reasonable agreement with the values recorded for R_{CO} in Table 16-6.

CRITICAL THINKING

13. Select a base resistor as needed to get $V_{CE} = 2$ V. You can use the data of Table 16-1 in calculating the required value of R_B.

14. Round off your calculated resistance to the nearest standard size. Then, build the circuit of Fig. 16-1 with the foregoing resistance. Measure V_{CE} to verify that it is reasonably close to 2 V.

COMPUTER (OPTIONAL)

15. Repeat Steps 1 to 14 using EWB or an equivalent circuit simulator. Do not record any new values. But make sure that you get reasonable agreement between the EWB measurements and the values recorded earlier.

16. If you are using the CD-ROM version of this book, click on the Assignments menu and select Chap. 6.

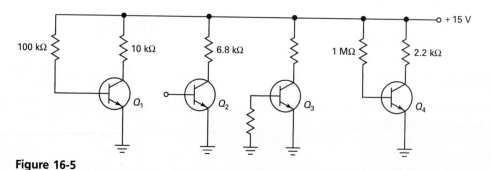

Figure 16-5

66

ADDITIONAL WORK (OPTIONAL)

17. In what region does the transistor of Fig. 16-1 operate? Prove your answer by showing your calculations here:

18. In what region does the transistor of Fig. 16-2 operate? Prove your answer by showing your calculations here:

19. In what region does the transistor of Fig. 16-3 operate? Prove your answer by showing your calculations here:

20. In what region does the transistor of Fig. 16-4 operate? Prove your answer by showing your calculations here:

Data for Experiment 16

TABLE 16-1. FIRST CIRCUIT

	V_{BE}	V_{CE}	Region
Calculated			
Measured			

TABLE 16-2. SECOND CIRCUIT

	V_{BE}	V_{CE}	Region
Calculated			
Measured			

TABLE 16-3. THIRD CIRCUIT

	V_{BE}	V_{CE}	Region
Calculated			
Measured			

TABLE 16-4. FOURTH CIRCUIT

	V_{BE}	V_{CE}	Region
Calculated			
Measured			

TABLE 16-5. FIFTH CIRCUIT

	V_{BE}	V_{CE}	Region
Q_1			
Q_2			
Q_3			
Q_4			

TABLE 16-6. TROUBLESHOOTING

	V_{BE}	V_{CE}	Region
R_{BO}			
R_{BG}			
R_{CO}			
R_{CS}			

Questions for Experiment 16

1. The base current in Fig. 16-1 is closest to: ( )
   (a) 0;   (b) 20 μA;   (c) 470 μA;   (d) 10 mA.
2. The collector current in Fig. 16-2 is closest to: ( )
   (a) 0;   (b) 20 μA;   (c) 470 μA;   (d) 10 mA.
3. The collector current in Fig. 16-3 is closest to: ( )
   (a) 0;   (b) 20 μA;   (c) 470 μA;   (d) 10 mA.
4. The collector voltage in Fig. 16-4 is closest to: ( )
   (a) 0;   (b) 2 V;   (c) 5 V;   (d) 10 V.
5. In Fig. 16-5, Q_1 operates in which region? ( )
   (a) active;   (b) saturation;   (c) cutoff;   (d) breakdown.
6. In Fig. 16-5, Q_4 operates in which region? ( )
   (a) active;   (b) saturation;   (c) cutoff;   (d) breakdown.
7. When the collector resistor of Fig. 16-1 is shorted, V_{CE} equals: ( )
   (a) 0;   (b) 2 V;   (c) 5 V;   (d) 10 V.
8. When the collector resistor of Fig. 16-1 is open, V_{CE} is closest to: ( )
   (a) 0;   (b) 2 V;   (c) 5 V;   (d) 10 V.
9. You are troubleshooting a circuit like Fig. 16-1 and have measured the collector-emitter voltage. How can you tell which of the regions the transistor is operating in?

10. Optional. Instructor's question.

70

Base Bias

A circuit like Fig. 17-1 is referred to as *base bias,* because it sets up a fixed base current. You can calculate the base current by applying Ohm's law to the total base resistance. This base current will remain constant when you replace transistors.

On the other hand, the collector current equals the current gain times the base current. Because of this, the collector current may have large variations from one transistor to the next. In other words, the Q point in a base-biased circuit is heavily dependent on the value of β_{dc}.

Required Reading

Chapter 7 (Secs. 7-1 to 7-5) of *Electronic Principles,* 6th ed.

Equipment

1 power supply: 15 V
3 transistors: 2N3904 (or almost any small-signal *npn* silicon transistor)
2 ½-W resistors: 2.2 kΩ and 22 kΩ
1 decade resistance box (or substitute a 1-MΩ potentiometer)

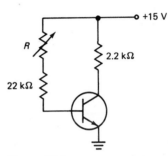

Figure 17-1

Procedure

1. The fixed-base-current circuit of Fig. 17-1 is not a stable biasing circuit, but it is a good way to measure β_{dc}.
2. Connect the circuit of Fig. 17-1 using one of the transistors.
3. Adjust R to get a V_{CE} of 1 V. Record the value of R in Table 17-1. (If R is a potentiometer instead of a decade box, you will have to disconnect it and measure its resistance.) In Fig. 17-1, notice the total base resistance R_B equals R plus 22 kΩ. Record the value of R_B in Table 17-1.
4. Calculate the values of β_{dc} and I_C. Record in Table 17-1.
5. Repeat Steps 2 through 4 for the second and third transistors.
6. With the values of Table 17-1, calculate the ideal and second-approximation values of I_E in Fig. 17-1. Record the I_E values in Table 17-2.

7. If a curve tracer or other transistor tester is available, measure the β_{dc} of each transistor for an I_C of approximately 5 mA. The values should be similar to the β_{dc} values of Table 17-1.

IDENTIFYING THE OPERATING REGION

8. In Fig. 17-1, calculate V_{BE}, $I_{C(sat)}$, $V_{CE(sat)}$, and $V_{CE(cutoff)}$. Record these values in Table 17-3.
9. Adjust R to 0 Ω and measure V_{BE}, I_C, and V_{CE}. Record these values in Table 17-4. Compare the measured values to the calculated values of Table 17-3. Decide which region the transistor is operating in. Record your answer in Table 17-4.
10. Disconnect one end of the 22-kΩ base resistor and measure V_{BE}, I_C, and V_{CE}. Record these values in Table 17-4. Decide which region the transistor is now operating in. Record your answer.
11. Adjust R until you measure V_{CE} as approximately 7.5 V. Measure V_{BE} and I_C. Record these values and the actual value of V_{CE} in Table 17-4. Which region is the transistor operating in? Record your answer.

APPLICATION (OPTIONAL)

12. A photocell can control a base-biased circuit to turn on different loads. As a demonstration, build the circuit of Fig. 17-2. The photocell should be exposed to incoming light.
13. Adjust R until you hear the buzzer and the LED lights up. Readjust R slowly until the buzzer and LED just turn off.
14. Now, cover the photocell to prevent light from reaching it. The buzzer and LED should come on.
15. Repeat Steps 13 and 14 for a few other photocells.

COMPUTER (OPTIONAL)

16. Repeat Steps 1 to 11 using EWB or an equivalent circuit simulator. Do not record any new values. But make sure that you get reasonable agreement between the EWB measurements and the values recorded earlier.

17. If you are using the CD-ROM version of this book, click on the Assignments menu and select Chap. 7.

ADDITIONAL WORK (OPTIONAL)

18. In this section, you will measure base and collector currents. Then, you will graph β_{dc} versus I_C.
19. Build the circuit of Fig. 17-3. Measure I_B and I_C for the minimum value of R. Record the values of I_B and I_C on a separate piece of paper.
20. Measure and record the foregoing currents for the maximum value of R.
21. Measure and record the foregoing currents for several intermediate values of R.
22. Calculate dc beta for each value of I_C. Then, graph dc beta versus the collector current. If you need additional points, repeat Step 21.

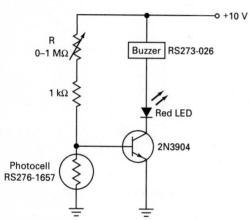

Figure 17-2

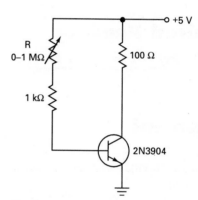

Figure 17-3

Data for Experiment 17

TABLE 17-1. β_{dc} VALUES

Transistor	R	R_B	β_{dc}	I_C
1				
2				
3				

TABLE 17-2. CALCULATIONS

Test	$I_{E(ideal)}$	$I_{E(second)}$
1		
2		
3		

TABLE 17-3. CALCULATIONS

V_{BE}	$I_{C(sat)}$	$V_{CE(sat)}$	$V_{CE(cutoff)}$

TABLE 17-4. MEASUREMENTS

Condition	V_{BE}	I_C	V_{CE}	Operating Region
$R = 0$				
22 kΩ open				
$V_{CE} \cong 7.5$ V				

Questions for Experiment 17

1. In Fig. 17-1, an increase in current gain causes an increase in: ( )
   (a) I_C; **(b)** V_{CE}; **(c)** I_B; **(d)** V_{CC}.
2. When R_B increases in a base-biased circuit, which of these increases? ( )
   (a) I_E; **(b)** V_{BB}; **(c)** I_C; **(d)** V_{CE}.
3. When R_C increases in a base-biased circuit, which of these decreases? ( )
   (a) I_E; **(b)** V_{BB}; **(c)** I_C; **(d)** V_{CE}.
4. When V_{BB} increases in a base-biased circuit, which of these decreases? ( )
   (a) I_B; **(b)** V_{BE}; **(c)** I_C; **(d)** V_{CE}.
5. When β_{dc} increases in a base-biased circuit, which of these remains constant? ( )
   (a) I_B; **(b)** I_E; **(c)** I_C; **(d)** V_{CE}.
6. The ideal and second-approximation values in Table 17-2 differ by approximately: ( )
   (a) 0.1%; **(b)** 1%; **(c)** 5%; **(d)** 10%.
7. If the base current is 10 μA in Fig. 17-1 and the collector voltage is 10 V, the current gain is closest to: ( )
   (a) 50; **(b)** 125; **(c)** 225; **(d)** 350.

8. If the collector voltage is 5 V in Fig. 17-1 and β_{dc} is 150, the base current is ( )
closest to:
   (a) 10 μA;   **(b)** 20 μA;   **(c)** 30 μA;   **(d)** μA.

9. Explain how you can tell when a transistor is operating in the cutoff region.

10. How can you tell whether a transistor is operating in the saturation region?

11. What about the active region? How can you identify it?

18

LED Drivers

The simplest way to use a transistor is as a switch, meaning that it operates at either saturation or cutoff but nowhere else along the load line. When saturated, a transistor appears as a closed switch between its collector and emitter terminals. When cut off, it is like an open switch. Because of the wide variation in β_{dc}, hard saturation is used with transistor switches. This means having enough base current to guarantee transistor saturation under all operating conditions. With small-signal transistors, hard saturation requires a base current of approximately one-tenth of the collector saturation current.

Another basic way to use the transistor is as a current source. In this case, the base resistor is omitted and the base supply voltage is connected directly to the base terminal. To set up the desired collector current, we use an emitter resistor. The emitter is bootstrapped to within one V_{BE} drop of the base voltage. Therefore, the collector current equals $(V_{BB} - V_{BE})$ divided by R_E. This fixed collector current then flows through the load, which is connected between the collector and positive supply voltage.

In this experiment you will build a transistor switch and a transistor current source. You will also have the opportunity to troubleshoot and design these basic transistor circuits.

Required Reading

Chapter 7 (Secs. 7-1 to 7-9) of *Electronic Principles*, 6th ed.

Equipment

2 power supplies: each source adjustable from 0 to 15 V
3 ½-W resistors: 220 Ω, 1 kΩ, 10 kΩ
1 LED: L53RD (or an equivalent red LED)
3 transistors: 2N3904
1 VOM (analog or digital multimeter)

Procedure

1. In Fig. 18-1, calculate I_B, I_C, and V_{CE}. Record your answers in Table 18-1.
2. Connect the transistor switch of Fig. 18-1. Measure and record the quantities listed in Table 18-1.
3. Repeat Steps 1 and 2 for the other transistors.

TRANSISTOR CURRENT SOURCE

4. In Fig. 18-2, calculate all quantities listed in Table 18-2.

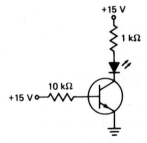

Figure 18-1

5. Connect the transistor current source of Fig. 18-2. Measure and record the quantities listed in Table 18-2.
6. Repeat Steps 4 and 5 for the other transistors.

TROUBLESHOOTING

7. In Fig. 18-1, assume the base resistor is open. Estimate and record the collector voltage in Table 18-3.
8. Repeat Step 7 for each of the troubles listed in Table 18-3.
9. Connect the circuit of Fig. 18-1 with each of the troubles listed in Table 18-3. Measure and record all listed quantities.
10. In Fig. 18-2, assume the emitter resistor is open. Estimate and record the voltages listed in Table 18-4.

75

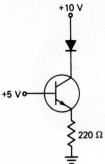

Figure 18-2

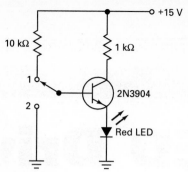

Figure 18-4

11. Repeat Step 10 for each of the troubles listed in Table 18-4.
12. Connect the circuit of Fig. 18-2 with each of the troubles listed in Table 18-4. Measure and record all listed quantities.

CRITICAL THINKING

13. Select a collector resistance in Fig. 18-1 to produce a collector current of approximately 30 mA. Calculate and record the quantities listed in Table 18-5.
14. Connect the circuit of Fig. 18-1 with your design value of collector resistance. Measure and record the quantities of Table 18-5.
15. Select an emitter resistance in Fig. 18-2 to get a collector current of approximately 30 mA. Calculate and record the quantities listed in Table 18-5.
16. Connect the circuit of Fig. 18-2 with your design value of emitter resistance. Measure and record all quantities in Table 18-5.

APPLICATION (OPTIONAL)

17. Measure the light and dark resistance of a few cadmium-sulfide photocells such as the Radio Shack 276-1657, an assorted package of five photoresistors,

devices whose resistance changes with the amount of incoming light.
18. Build the circuit of Fig. 18-3. Adjust the supply voltage until the LED is dimly lit. Cover the photocell and notice how the LED either goes out or becomes less bright. Repeat this demonstration with a few other photocells.

COMPUTER (OPTIONAL)

19. Repeat Steps 1 to 16 using EWB or an equivalent circuit simulator. Do not record any new values. But make sure that you get reasonable agreement between the EWB measurements and the values recorded earlier.
20. If you are using the CD-ROM version of this book, click on the Assignments menu and select Chap. 7.

ADDITIONAL WORK (OPTIONAL)

21. Connect the circuit of Fig. 18-4. Record whether the LED is on or off for each position of the switch.
22. Connect the circuit of Fig. 18-5. For each switch position, compare the on-off state of the LED to the recorded values of Step 21. What did you learn?

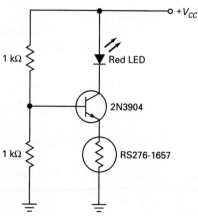

Figure 18-3

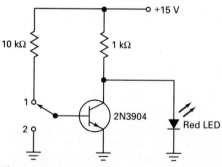

Figure 18-5

Data for Experiment 18

TABLE 18-1. TRANSISTOR SWITCH

Transistor	Calculated			Measured		
	I_B	I_C	V_{CE}	I_B	I_C	V_{CE}
1						
2						
3						

TABLE 18-2. TRANSISTOR CURRENT SOURCE

Transistor	Calculated			Measured		
	V_E	I_C	V_{CE}	V_E	I_C	V_{CE}
1						
2						
3						

TABLE 18-3. TROUBLESHOOTING THE TRANSISTOR SWITCH

Trouble	Estimated V_C	Measured V_C
Open 10 kΩ		
Open 1 kΩ		
Shorted collector-emitter		
Open collector-emitter		

TABLE 18-4. TROUBLESHOOTING THE TRANSISTOR CURRENT SOURCE

Trouble	Estimated		Measured	
	V_C	V_E	V_C	V_E
Open 220 Ω				
Shorted collector-emitter				
Open collector-emitter				

TABLE 18-5. CRITICAL THINKING

Transistor	Calculated			Measured	
	R	V_E	I_C	V_E	I_C
Switch					
Current source					

Questions for Experiment 18

1. In Fig. 18-1, the ratio of collector current to base current is closest to: ( )
   (a) 1;   (b) 10;   (c) 100;   (d) 300.
2. The measured V_{CE} entries of Table 18-1 indicate that collector voltage is approx-   ( )
   imately;
   (a) 0;   (b) 2 V;   (c) 4 V;   (d) 8 V.
3. In the transistor current source of Fig. 18-2, the emitter voltage is closest to: ( )
   (a) 0.7 V;   (b) 4.3 V;   (c) 5 V;   (d) 10 V.
4. When a transistor is in hard saturation, its collector-emitter terminals appear ap-   ( )
   proximately:
   (a) shorted;   (b) open;   (c) in the active region;   (d) cut off.
5. With a transistor current source; the emitter is bootstrapped to within one V_{BE} drop   ( )
   of the:
   (a) base voltage;   (b) emitter voltage;   (c) collector voltage;   (d) collector current.
6. What are some of the differences between a transistor switch and a transistor current source?

TROUBLESHOOTING

7. While troubleshooting a transistor switch like Fig. 18-1, you notice that the collector voltage is always zero. If the LED is lit, what is the most likely trouble?

8. Explain the measured values for collector and emitter voltage when the emitter resistor was
   open in Fig. 18-2.

DESIGN

9. Why is hard saturation used with a transistor switch?

10. Optional. Instructor's question.

Setting Up a Stable Q Point

If you want a stable *Q* point, you will have to use either voltage-divider bias or two-supply emitter bias. With either of these stable biasing methods, the effects of h_{FE} variations are virtually eliminated. Voltage-divider bias requires only a single power supply. This type of bias is also called universal bias, an indication of its popularity. When two supplies are available, two-supply emitter bias can provide as stable a *Q* point as voltage-divider bias.

In this experiment you will connect both types of bias and verify the stable *Q* points discussed in your textbook.

Required Reading

Chapter 8 (Secs. 8-1 to 8-4) of *Electronic Principles,* 6th ed.

Equipment

2 power supplies: 15 V
3 transistors: 2N3904 (or equivalent)
5 ½-W resistors: 1 kΩ, 2.2 kΩ, 3.9 kΩ, 8.2 kΩ, 10 kΩ
1 VOM (analog or digital multimeter)

Procedure

VOLTAGE-DIVIDER BIAS

1. In Fig. 19-1, calculate V_B, V_E, and V_C. Record your answers in Table 19-1.

2. Connect the circuit of Fig. 19-1. Measure and record the quantities listed in Table 19-1.
3. Repeat Steps 1 and 2 for the other transistors.

EMITTER BIAS

4. In Fig. 19-2, calculate V_B, V_E, and V_C. Record your answers in Table 19-2.
5. Connect the emitter-biased circuit of Fig. 19-2. Measure and record the quantities of Table 19-2.
6. Repeat Steps 4 and 5 for the other transistors.

TROUBLESHOOTING

7. In Fig. 19-1, assume that R_1 is open. Estimate and record the collector voltage V_C in Table 19-3.
8. Repeat Step 7 for the other troubles listed in Table 19-3. Connect the circuit of Fig. 19-1 with each trou-

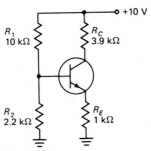

Figure 19-1

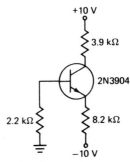

Figure 19-2

ble listed in Table 19-3. Measure and record the collector voltage.

CRITICAL THINKING

9. Design a stiff voltage-divider biased circuit to meet the following specifications: $V_{CC} = 15$ V, $I_C = 2$ mA, and $V_C = 7.5$ V. You may assume an h_{FE} of 200. Calculate and record the quantities listed in Table 19-4.
10. Connect your design. Measure and record the quantities of Table 19-4.

COMPUTER (OPTIONAL)

11. Repeat Steps 1 to 10 using EWB or an equivalent circuit simulator. Do not record any new values. But make sure that you get reasonable agreement between the EWB measurements and the values recorded earlier.
12. If you are using the CD-ROM version of this book, click on the Assignments menu and select Chap. 8.

ADDITIONAL WORK (OPTIONAL)

13. Assume $\beta_{dc} = 100$ for the emitter-feedback biased circuit of Fig. 19-3. Calculate and record V_B, V_E, and V_C on a separate piece of paper. (Use a table similar to Table 19-1 for your data.)
14. Build the circuit of Fig. 19-3. Measure and record V_B, V_E, and V_C.
15. Compare the measured values to the calculated values. What does this tell you about β_{dc}?
16. Repeat Steps 13 to 15 for the collector-feedback biased circuit of Fig. 19-4.

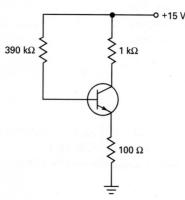

Figure 19-3

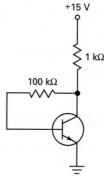

Figure 19-4

Data for Experiment 19

TABLE 19-1. VOLTAGE-DIVIDER BIAS

Transistor	Calculated V_B	V_E	V_C	Measured V_B	V_E	V_C
1						
2						
3						

TABLE 19-2. EMITTER BIAS

Transistor	Calculated V_B	V_E	V_C	Measured V_B	V_E	V_C
1						
2						
3						

TABLE 19-3. TROUBLESHOOTING

Trouble	Estimated V_C	Measured V_C
Open R_1		
Shorted R_1		
Open R_2		
Shorted R_2		
Open R_C		
Shorted R_C		
Open R_E		
Shorted R_E		
Open collector-emitter		
Shorted collector-emitter		

TABLE 19-4. CRITICAL THINKING

Values: $R_1 =$ _____ ; $R_2 =$ _____ ; $R_C =$ _____ ; $R_E =$ _____

Transistor	Calculated V_C	Measured V_C
1		
2		
3		

Questions for Experiment 19

1. Ideally, the voltage divider of Fig. 19-1 produces which of the following base ( )
   voltages:
   (a) 0 V;   (b) 1.1 V;   (c) 1.8 V;   (d) 6.03 V.
2. The measured emitter voltage of Fig. 19-1 was closest to: (b)
   (a) 0 V;   (b) 1.1 V;   (c) 1.8 V;   (d) 6.03 V.
3. The measured collector voltage of Fig. 19-1 was closest to: ( )
   (a) 0 V;   (b) 1.1 V;   (c) 1.8 V;   (d) 6.03 V.
4. The base voltage measured in Fig. 19-2 was: (d)
   (a) 0 V;   (b) slightly positive;   (c) slightly negative;   (d) −0.7 V.
5. With both voltage-divider bias and emitter bias, the measured collector voltage was ( )
   approximately:
   (a) constant;   (b) negative;   (c) unstable;   (d) one V_{BE} drop less than
   the base voltage.
6. What did you learn about the Q point of a circuit that uses voltage-divider bias or emitter
   bias?

TROUBLESHOOTING

7. Name all the troubles you found that produced a collector voltage of 10 V.

8. What collector voltage did you measure with a shorted collector-emitter? Explain why this
   value occurred.

CRITICAL THINKING

9. Compare the measured V_C with the calculated V_C in Table 19-4. Explain why the measured
   and calculated values differ.

10. Optional. Instructor's question.

21

Transistor Bias

Before we can use a transistor to amplify an ac signal, we have to set up a quiescent (Q) point of operation, typically near the middle of the dc load line. Then, the incoming ac signal can produce fluctuations above and below this Q point. The three most primitive forms of bias are base bias, emitter-feedback bias, and collector-feedback bias. As you know, these are not the best ways to bias a transistor if you want a stable Q point. Nevertheless, you may occasionally see these biasing methods used with small-signal amplifiers. In this experiment you will connect all three types of bias to verify the operation as discussed in your textbook.

The most common transistor troubles are the collector-emitter short and collecter-emitter open. We will simulate the collector-emitter short by putting a jumper between the collector, base, and emitter; this is equivalent to shorting both diodes. We will simulate a collector-emitter open by removing the transistor from the circuit; this is equivalent to opening both diodes.

Required Reading

Chapter 8 (Secs. 8-5 and 8-6) of *Electronic Principles*, 6th ed.

Equipment

1 power supply: 15 V
3 transistors: 2N3904 (or equivalent)
7 ½-W resistors: 100 Ω, 680 Ω, 820 Ω, 1 kΩ, 220 kΩ, 270 kΩ, 470 kΩ
1 VOM (analog or digital multimeter)

Procedure

BASE BIAS

1. Refer to the data sheet of a 2N3904 in the Appendix. Notice that the dc current gain h_{FE} has a minimum value of 100 and a maximum value of 300 for an I_C of 10 mA. The typical value is not listed. For this experiment we will assume the typical value is 200.
2. In Fig. 21-1, use the typical h_{FE} to calculate I_B, I_C, and V_C. Record your answers in Table 21-1.
3. Connect the circuit of Fig. 21-1. Measure and record the quantities listed in Table 21-1.
4. Repeat Steps 2 and 3 for the other transistors.

Figure 21-1

EMITTER-FEEDBACK BIAS

5. In Fig. 21-2, use the typical h_{FE} to calculate I_C, V_C, and V_E. Record your answers in Table 21-2.
6. Connect the emitter-feedback bias of Fig. 21-2. Measure and record the quantities of Table 21-2.
7. Repeat Steps 5 and 6 for the other transistors.

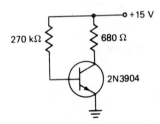

Figure 21-2

COLLECTOR-FEEDBACK BIAS

8. In Fig. 21-3, use the typical h_{FE} to calculate and record the quantities of Table 21-3.
9. Connect the circuit of Fig. 21-3. Measure and record all quantities listed in Table 21-3.
10. Repeat Steps 8 and 9 for the other transistors.

TROUBLESHOOTING

11. In Fig. 21-3, assume that the base resistor is open. Estimate and record the collector voltage V_C in Table 21-4.
12. Repeat Step 11 for the other troubles listed in Table 21-4.
13. Connect the circuit of Fig. 21-3 with each trouble listed in Table 21-4. Measure and record the collector voltage.

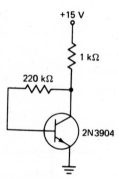

Figure 21-3

CRITICAL THINKING

14. Design a collector-feedback biased circuit with a 2N3904 to meet the following specifications: $V_{CC} = 10$ V and $I_C = 2$ mA. *Hint:* Look at the 2N3904 data sheet. Calculate and record the quantities listed in Table 21-5.
15. Connect your design. Measure and record the quantities of Table 21-5.

COMPUTER (OPTIONAL)

16. Repeat Steps 1 to 15 using EWB or an equivalent circuit simulator. Do not record any new values. But make sure that you get reasonable agreement between the EWB measurements and the values recorded earlier.
17. If you are using the CD-ROM version of this book, click on the Assignments menu and select Chap. 8.

ADDITIONAL WORK (OPTIONAL)

18. Connect the circuit of Fig. 21-1 using a 2N3906 and a collector supply voltage of -15 V. Measure V_B, V_E, and V_C. Record the value on a separate piece of paper.
19. Connect the circuit of Fig. 21-2 using a 2N3906 and a collector supply voltage of -15 V. Measure and record V_B, V_E, and V_C.
20. Connect the circuit of Fig. 21-3 using a 2N3906 and a collector supply voltage of -15 V. Measure and record V_B, V_E, and V_C.
21. Compare the recorded values of Steps 18 through 20 with the values recorded in Tables 21-1 through 21-3. What did you learn about *npn* and *pnp* biasing circuits?

88

Data for Experiment 21

TABLE 21-1. BASE BIAS

Transistor	Calculated			Measured		
	I_B	I_C	V_C	I_B	I_C	V_C
1						
2						
3						

TABLE 21-2. EMITTER-FEEDBACK BIAS

Transistor	Calculated			Measured		
	I_C	V_C	V_E	I_C	V_C	V_E
1						
2						
3						

TABLE 21-3. COLLECTOR-FEEDBACK BIAS

Transistor	Calculated			Measured		
	V_B	I_C	V_C	V_B	I_C	V_C
1						
2						
3						

TABLE 21-4. TROUBLESHOOTING

Trouble	Estimated V_C	Measured V_C
Open 200 kΩ		
Shorted 200 kΩ		
Open 1 kΩ		
Shorted 1 kΩ		
Open collector-emitter		
Shorted collector-emitter		

TABLE 21-5. CRITICAL THINKING: $R_B =$ _____ ; $R_C =$ _____

Transistor	Calculated V_C	Measured V_C
1		
2		
3		

Questions for Experiment 21

1. Base bias has an unstable Q point because of the variation in: ( )
   (a) base current;   (b) V_{BE};   (c) base resistance;   (d) h_{FE}.

2. When the collector current increases in a base-biased circuit, the collector voltage: ( )
   (a) increases;   (b) stays the same;   (c) decreases.

3. In Fig. 21-2, the collector saturation current has a value of approximately: ( )
   (a) 5 mA;   (b) 10 mA;   (c) 15 mA;   (d) 20 mA.

4. The measured data of Table 21-3 show that the V_{BE} drop was closest to: ( )
   (a) 0;   (b) 0.3 V;   (c) 0.7 V;   (d) 7.85 V.

5. Of the three circuits tested, which had the most stable Q point?   ( )
   (a) base bias;   (b) emitter-feedback bias;   (c) collector-feedback bias
   (d) voltage-divider bias.

6. Briefly discuss the Q point for the three circuits tested.

TROUBLESHOOTING

7. Assume that you are troubleshooting a circuit like Fig. 21-3. If you measure a collector voltage V_C of 15 V, what are three possible troubles?

8. Name two possible troubles in Fig. 21-3 that would produce a collector voltage of zero.

CRITICAL THINKING

9. Explain how you calculated your design values in Table 21-5.

10. Optional. Instructor's question.

Coupling and Bypass Capacitors

Capacitive reactance decreases as frequency increases. Because of this, a capacitor has a large impedance at low frequencies and a small impedance at high frequencies. As an approximation, we can say that a capacitor is a dc open and an ac short. When used in amplifiers, capacitors can couple the signal from one active node to another, or they can bypass the signal from an active node to ground.

The cutoff frequency is the frequency where $X_c = R$. From this, we can derive the following equation for the cutoff frequency:

$$f_c = \frac{1}{2\pi RC}$$

In Fig. 22-1a, $R = 68$ kΩ $\|$ 100 kΩ. In Fig. 22-1b, $R = 22$ kΩ $\|$ 22 kΩ. At the cutoff frequency, the output voltage is 0.707 times the input voltage.

Required Reading

Chapter 9 (Secs. 9-1 and 9-2) of *Electronic Principles*, 6th ed.

Equipment

1 audio generator
4 ½-W resistors: two 22 kΩ, 68 kΩ, 100 kΩ
1 capacitor: 0.022 μF
1 oscilloscope

Procedure

1. Calculate the cutoff frequency in Fig. 22-1a. Fill in the values of f_c, $10f_c$, and $0.1f_c$ in Table 22-1 under f.
2. Connect the circuit of Fig. 22-1a.
3. Adjust the audio generator to get a frequency of f_c and an input voltage v_{in} of 1 V peak-to-peak on the oscilloscope.
4. Measure the output voltage v_{out} and record in Table 22-1.
5. Change the frequency to $10f_c$ and readjust to get a v_{in} of 1 V pp. Measure v_{out} and record in Table 22-1.
6. Change the frequency to $0.1f_c$ and check that v_{in} is 1 V pp. Measure and record v_{out}.

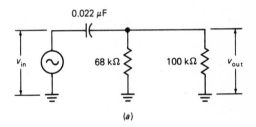

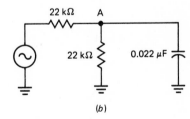

Figure 22-1

7. Calculate the cutoff frequency in Fig. 22-1b. Fill in the values of f_c, $10f_c$, and $0.1f_c$ in Table 22-2.
8. Connect the circuit of Fig. 22-1b without the capacitor. Adjust the frequency to f_c.
9. Set the signal level to 1 V pp across the lower 22-kΩ resistor.
10. Connect the capacitor between point A and ground. Then measure and record v_A.

11. Remove the capacitor and change the frequency to $10f_c$. Then repeat Steps 9 and 10.
12. Remove the capacitor and change the frequency to $0.1f_c$. Then repeat Steps 9 and 10.

CRITICAL THINKING (OPTIONAL)

13. This may require reading ahead in your textbook if you are not familiar with Bode plots. Construct an idealized Bode plot of the data in Tables 22-1 and 22-2 on two-cycle, semilogarithmic graph paper. If you are not familiar with a Bode plot, you will find information in Chap. 16 of *Electronic Principles*. Use either two different colors or two different patterns for the lines on one set of axes. Attach your graph to the report for this experiment.

COMPUTER (OPTIONAL)

14. Repeat Steps 1 to 12 using EWB or an equivalent circuit simulator. Do not record any new values. But make sure that you get reasonable agreement between the EWB measurements and the values recorded earlier.

15. If you are using the CD-ROM version of this book, click on the Assignments menu and select Chap. 9.

ADDITIONAL WORK (OPTIONAL)

16. Build the circuit of Fig. 22-1a with a small 8-Ω speaker connected across the 100 kΩ. Replace the 0.022 μF with 0.47 μF. Increase the output of the audio generator until you can hear an audio tone. Vary the frequency and notice how the tone changes pitch. Leave the output level of the audio generator fixed during the next step.
17. Build the circuit of Fig. 22-1b with a small 8-Ω speaker across the 0.022 μF. With the output level of the audio generator set the same as in the preceding step, what do you hear?
18. What did you learn from Steps 16 and 17? Why does this happen?

92

Data for Experiment 22

TABLE 22-1. COUPLING CAPACITOR

	f	v_{out}
f_c		
$10f_c$		
$0.1f_c$		

TABLE 22-2. BYPASS CAPACITOR

	f	v_A
f_c		
$10f_c$		
$0.1f_c$		

Questions for Experiment 22

1. A coupling capacitor ideally looks like a dc: ( )
   (a) open and ac open;   (b) open and ac short;   (c) short and ac open;
   (d) short and ac short.

2. A small Thevenin resistance means the bypass capacitor must be: ( )
   (a) small;   (b) large;   (c) unaffected;   (d) open.

3. The value of f_c in Table 22-1 is closest to: ( )
   (a) 100 Hz;   (b) 180 Hz;   (c) 1 kHz;   (d) 5 kHz.

4. The value of f_c in Table 22-2 is closest to: ( )
   (a) 100 Hz;   (b) 500 Hz;   (c) 650 Hz;   (d) 1 kHz.

5. If we want an f_c of 18 Hz in Fig. 22-1a, we have to change the capacitor to   ( )
   approximately:
   (a) 0.1 μF;   (b) 0.2 μF;   (c) 0.5 μF;   (d) 1 μF.

6. In Table 22-1, the output voltage at f_c is closest to: ( )
   (a) 0.707 V;   (b) 0.9 V;   (c) 0.99 V;   (d) 1 V.

7. In Table 22-2, the output voltage at f_c is closest to: ( )
   (a) 0.01 V;   (b) 0.15 V;   (c) 0.707 V;   (d) 1 V.

8. When the input voltage is 1 V in Fig. 22-1a, the output voltage at $10f_c$ is closest to: ( )
   (a) 0;   (b) 0.707 V;   (c) 0.9 V;   (d) 1 V.

9. Optional. Instructor's question.

10. Optional. Instructor's question.

The CE Amplifier

After the transistor of a CE amplifier has been biased with its Q point near the middle of the dc load line, you can couple a small ac signal into the base. This produces an amplified ac signal at the collector. In this experiment you will build a CE amplifier and measure its voltage gain, as well as looking at the dc and ac waveforms throughout the circuit.

Required Reading

Chapter 10 (Sec. 10-1) of *Electronic Principles*, 6th ed.

Equipment

1   audio generator
1   power supply: 10 V
3   transistors: 2N3904 (or equivalent)
4   ½-W resistors: 1 kΩ, 2.2 kΩ, 3.9 kΩ, 10 kΩ
2   capacitors: 1 μF, 470 μF (10-V rating or better)
1   oscilloscope

Procedure

DC AND AC VOLTAGES

1. In Fig. 23-1, calculate the dc voltage at the base, emitter, and collector. Record your answers in Table 23-1.
2. Calculate and record the peak-to-peak ac voltages at the base, emitter, and collector.

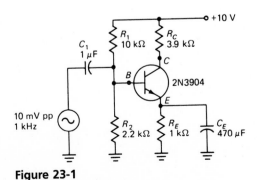

Figure 23-1

3. Connect the circuit. Adjust the signal generator to get an input signal of 10 mV peak-to-peak at 1 kHz.
4. Look at the base, emitter, and collector. At each point, use the oscilloscope on dc input to measure dc voltage and on ac input to measure the ac peak-to-peak voltage. Record all voltages in Table 23-1.
5. With the oscilloscope on dc input, you should see waveforms like those of Fig. 9-7 in your textbook. This confirms that the total voltages are the sum of dc and ac components.

PHASE INVERSION

6. If you are using a dual-trace or dual-beam oscilloscope, look at the base signal with one input and the collector signal with the other input. Also use the collector signal to drive the external trigger of the oscilloscope. (If in doubt about how to use the external trigger, ask the instructor.) Notice that the collector signal is 180° out of phase with the base signal.
7. If you are using a single-trace oscilloscope, externally trigger the oscilloscope with the collector signal. (If in doubt, ask the instructor about this.) Look first at the base signal, then at the collector signal. Notice that the signals are 180° out of phase.

VOLTAGE GAIN

8. In Fig. 23-1, use Eq. (9-10) in your textbook to calculate the ideal emitter resistance r'_e. Use Eq. (10-3) to calculate the voltage gain A. Record your answers in Table 23-2.
9. Connect the circuit with any of the three transistors. Measure and record the input and output ac voltages.
10. Calculate the actual voltage gain using the v_{out} and v_{in} measured in Step 9. Next calculate r'_e using the

ratio R_C/A. Record your experimental A and r'_e in Table 23-2.

11. Repeat Steps 8 to 10 for the other transistors.

TROUBLESHOOTING

12. In Fig. 23-1, assume that C_1 is open. Estimate the peak-to-peak ac voltage at the base, emitter, and collector. Record in Table 23-3.
13. Repeat Step 12 for each trouble listed in Table 23-3.
14. Connect the circuit with each trouble. Measure and record the ac voltages.

CRITICAL THINKING

15. Select a value of collector resistance in Fig. 23-1 to produce a theoretical voltage gain of 100. Using the nearest standard resistance, calculate and record the quantities of Table 23-4.
16. Connect the circuit with your design value of R_C. Measure and record the quantities listed in Table 23-4.
17. Repeat Step 16 for the other transistors.

COMPUTER (OPTIONAL)

18. Repeat Steps 1 to 17 using EWB or an equivalent circuit simulator. Do not record any new values. But make sure that you get reasonable agreement between the EWB measurements and the values recorded earlier.
19. If you are using the CD-ROM version of this book, click on the Assignments menu and select Chap. 10.

ADDITIONAL WORK (OPTIONAL)

20. Have another student insert one of the following troubles into the circuit: open or short any resistor, open any capacitor, open any connecting wire, short the BE, CB, or CE terminals of the transistor. Use only voltage readings of a DMM or an oscilloscope to troubleshoot.
21. Repeat Step 20 several times until you are confident that you can troubleshoot the circuit for various troubles.

Data for Experiment 23

TABLE 23-1. CE AMPLIFIER

	Calculated			Measured		
	B	E	C	B	E	C
dc						
ac						

TABLE 23-2. VOLTAGE GAIN

Transistor	Calculated		Measured		Experimental	
	r_e'	A	v_{in}	v_{out}	A	r_e'
1						
2						
3						

TABLE 23-3. TROUBLESHOOTING

Trouble	Estimated			Measured		
	v_b	v_e	v_c	v_b	v_e	v_c
Open C_1						
Open R_2						
Open R_E						

TABLE 23-4. CRITICAL THINKING

Transistor	Calculated			Measured		
	r_e'	R_C	A	v_{in}	v_{out}	A
1						
2						
3						

Questions for Experiment 23

1. The CE amplifier of Fig. 23-1 has a theoretical r_e' of: ( )
   (a) 22.7 Ω;   (b) 1 kΩ;   (c) 3.6 kΩ;   (d) 10 kΩ.
2. Ideally, the CE amplifier of Fig. 23-1 has a voltage gain of approximately: ( )
   (a) 1;   (b) 3.6;   (c) 4.54;   (d) 159.
3. The emitter of Fig. 23-1 had little or no ac signal because of: ( )
   (a) the emitter resistor;   (b) the input coupling capacitor;   (c) the emitter bypass capacitor;   (d) the weak base signal.
4. The voltage at collector was closest to: ( )
   (a) 6 V dc and 10 mV ac;   (b) 1.8 V dc and 1.6 V ac;   (c) 1.1 V dc and 10 mV ac;   (d) 6 V dc and 1.6 V ac.

5. The dc bias of the transistor is undisturbed by the dc resistance of the signal generator because the input coupling capacitor: ( )
(**a**) blocks dc; (**b**) transmits ac; (**c**) blocks ac; (**d**) transmits dc.

6. As briefly as possible, explain how the circuit of this experiment amplifies the signal.

TROUBLESHOOTING

7. What happens to the dc voltages of the circuit when the coupling capacitor is open? What happens to the ac voltages?

8. Explain the measured voltages you got with an open R_E.

CRITICAL THINKING

9. Explain how you selected the value of load resistance.

10. Optional. Instructor's question.

Other CE Amplifiers

Because of the input impedance of a CE amplifier, some of the ac signal may be dropped across the source impedance. Furthermore, the Thevenin equivalent of the amplifier output is an ac generator in series with the output impedance of the amplifier. When a load resistance is connected to the amplifier, some of the ac signal is dropped across the output impedance.

One way to increase the input impedance is to use a swamping resistor in the emitter circuit. This also stabilizes the voltage gain against changes in r'_e. Because the swamping resistor reduces the voltage gain, it may be necessary to cascade two swamped amplifiers to get the same voltage gain as a single unswamped stage.

Required Reading

Chapter 10 (Secs. 10-1 to 10-2) of *Electronic Principles*, 6th ed.

Equipment

1   audio generator
1   power supply: 10 V
3   transistors: 2N3904 (or equivalent)
7   ½-W resistors: two 1 kΩ, 1.5 kΩ, 2.2 kΩ, 3.9 kΩ, 10 kΩ, 51 kΩ
3   capacitors: two 1 μF, 470 μF (10-V rating or better)
1   oscilloscope

Procedure

CE AMPLIFIER WITH SOURCE AND LOAD RESISTANCES

1. In Fig. 24-1, assume h_{fe} (same as β) is 150. Calculate the input impedance of the stage. Also calculate the peak-to-peak base voltage and the peak-to-peak collector voltage. Record your answers in Table 24-1.
2. Connect the circuit. Adjust the signal generator to get a source signal of 20 mV peak-to-peak at 1 kHz. (Measure this between the left end of the source resistance and ground.)
3. Look at the base and collector. At each point, use the oscilloscope on dc input to verify that the waveforms

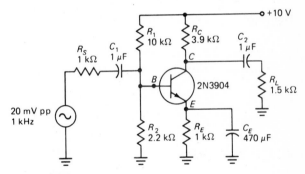

Figure 24-1

are the sum of dc and ac components. Also look at the emitter. Because of the bypass capacitor, the emitter should have only a dc component.

4. With the oscilloscope on ac input, measure the peak-to-peak voltage at the base and collector. Record your data in Table 24-1.
5. Repeat Step 4 for the two other transistors.

SWAMPED AMPLIFIER

6. In Fig. 24-2, assume an h_{fe} of 150 and calculate the input impedance of the stage. Also calculate the peak-to-peak voltage at the base and collector. Record your answers in Table 24-2.
7. Connect the circuit. Measure and record the peak-to-peak voltage at the base and collector.
8. Repeat Steps 6 and 7 for the other two transistors.

TROUBLESHOOTING

9. In Fig. 24-2, assume that C_E is open. Estimate the peak-to-peak ac voltage at the base, emitter, and collector. Record in Table 24-3.
10. Repeat Step 9 for each trouble listed in Table 24-3.
11. Connect the circuit with each trouble. Measure and record the ac voltages.

CRITICAL THINKING

12. Select a value of swamping resistance in Fig. 24-2 to produce an unloaded voltage gain of 10 from the base to the collector. Using the nearest standard resistance, calculate and record the quantities of Table 24-4.
13. Connect the circuit with your design value of r_E. Measure and record the quantities listed in Table 24-4.
14. Repeat Step 13 for the other transistors.

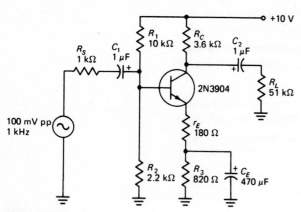

Figure 24-2

COMPUTER (OPTIONAL)

15. Repeat Steps 1 to 14 using EWB or an equivalent circuit simulator. Do not record any new values. But make sure that you get reasonable agreement between the EWB measurements and the values recorded earlier.
16. If you are using the CD-ROM version of this book, click on the Assignments menu and select Chap. 10.

ADDITIONAL WORK (OPTIONAL)

17. Have another student insert one of the following troubles into the circuit of Fig. 24-2: open or short any resistor, open any capacitor, open any connecting wire, short the BE, CB, or CE terminals of the transistor. Use only voltage readings of a DMM or an oscilloscope to troubleshoot.
18. Repeat Step 17 several times until you are confident that you can troubleshoot the circuit for various troubles.

100

Data for Experiment 24

TABLE 24-1. CE AMPLIFIER WITH SOURCE AND LOAD RESISTANCES

Transistor	Z_{in}	Calculated v_b	v_c	Measured v_b	v_c
1					
2					
3					

TABLE 24-2. SWAMPED AMPLIFIER

Transistor	Z_{in}	Calculated v_b	v_c	Measured v_b	v_c
1					
2					
3					

TABLE 24-3. TROUBLESHOOTING

Trouble	Estimated v_b	v_e	v_c	Measured v_b	v_e	v_c
Open C_E						
Shorted C_E						
Open collector-emitter						
Shorted collector-emitter						
Open C_2						
Shorted C_2						

TABLE 24-4. CRITICAL THINKING

Transistor	r_e	Calculated v_b	v_c	Measured v_b	v_c
1					
2					
3					

Questions for Experiment 24

1. The calculated base voltage in Table 24-1 is approximately: ( )
   (a) 10.8 mV; **(b)** 20 mV; **(c)** 250 mV; **(d)** 500 mV.
2. The calculated collector voltage in Table 24-1 was closest to: ( )
   (a) 10.8 mV; **(b)** 20 mV; **(c)** 250 mV; **(d)** 500 mV.

3. In Table 24-2, the measured base voltage was closest to: ( )
    (a) 12 mV;   **(b)** 20 mV;   **(c)** 63 mV;   **(d)** 1 V.

4. The voltage gain from base to collector in the swamped amplifier was closest to: ( )
    (a) 1;   **(b)** 5;   **(c)** 10;   **(d)** 15.

5. Compared with the CE amplifier, the swamped amplifier had a: ( )
    (a) lower input impedance;   **(b)** higher output impedance;   **(c)** lower
voltage gain;   **(d)** lower ac collector voltage.

6. Explain why an amplifier with a swamping resistor has a more stable voltage gain than one without.

TROUBLESHOOTING

7. What happens to the voltage gain of an amplifier when the emitter bypass capacitor is open? Explain why this happens.

8. Explain what happens when the emitter bypass capacitor is shorted.

CRITICAL THINKING

9. How did you arrive at the value of the swamping resistor?

10. Optional. Instructor's question.

Cascaded CE Stages

The amplified signal out of a CE stage can be used as the input to another CE stage. In this way, we can build a multistage amplifier with very large voltage gain. Because a CE stage has an input impedance, there is a loading effect on the preceding stage. In other words, the loaded voltage gain is less than the unloaded voltage gain. In this experiment, you will build a two-stage amplifier using swamped stages to stabilize the overall voltage gain.

Required Reading

Chapter 10 (Secs. 10-3 to 10-6) of *Electronic Principles*, 6th ed.

Equipment

- 1 audio generator
- 1 power supply: 10 V
- 2 transistors: 2N3904 (or equivalent)
- 12 ½-W resistors: two 68 Ω, three 1 kΩ, one 1.2 kΩ, two 2.2 kΩ, two 3.9 kΩ, two 10 kΩ
- 5 capacitors: three 1 μF, two 47 μF (10-V rating or better)
- 1 VOM (analog or digital multimeter)
- 1 oscilloscope

Procedure

CALCULATIONS

1. In Fig. 25-1, calculate the dc voltages at the base, emitter, and collector of each stage. Record your answers in Table 25-1.
2. Next, examine the 2N3904 data sheet in the Appendix. Read the typical value of h_{fe}. Record this value at the top of Table 25-2.
3. Calculate the peak-to-peak ac voltage at the base, emitter, and collector of each stage (Fig. 25-1) using the h_{fe} of Step 2. Record all ac voltages in Table 25-2.

TESTS

4. Connect the two-stage amplifier of Fig. 25-1.
5. Measure the dc voltage at the base, emitter, and collector of each stage. Record your data in Table 25-1. Within the tolerance of the resistors being used, the measured voltages should agree with your calculated voltages.
6. Measure the peak-to-peak ac voltage at the base, emitter, and collector of each stage. Record your data in Table 25-2. These measured ac voltages should agree with your calculated values.

LOADING EFFECTS

7. Open the coupling capacitor between the first and second stage. Look at the ac voltage on the first collector. Reconnect the coupling capacitor and notice that the ac signal decreases significantly. Do not continue until you understand and can explain why the signal decreases.
8. Open the coupling capacitor between the second stage and the load resistor. Look at the ac voltage on the second collector. Reconnect the coupling capacitor and notice the decrease in signal strength. Once more, you should be able to explain why this happens.

TROUBLESHOOTING

9. In Fig. 25-1, assume that C_4 is open. Does this produce a trouble in the first or second stage? Record your answer (1 or 2) in Table 25-3.

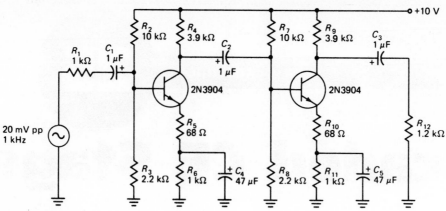

Figure 25-1

10. Insert the foregoing trouble in your circuit. You will be measuring dc and ac voltages. Before you measure each voltage, make a ballpark estimate of its value. Then when you measure the voltage, you will know whether or not it is a clue to the trouble.

11. Estimate each voltage in Table 25-3 for the stage with the trouble. Measure and record the voltage.

12. Repeat Steps 9 to 11 for each of the troubles listed in Table 25-3.

CRITICAL THINKING

13. Select a swamping resistor for the second stage to get an overall voltage gain of approximately 75. Record your design value at the top of Table 25-4.

14. Change the swamping resistor to your design value. Measure and record the voltage gain of the first stage (base to collector). Then measure and record the voltage gain of the second stage.

15. Measure and record the overall voltage gain (first base to second collector).

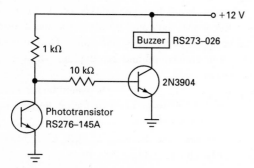

Figure 25-2

APPLICATION (OPTIONAL)

16. Connect the circuit shown in Fig. 25-2 using a phototransistor in the first stage and a magnetic buzzer in the second stage.

17. After power is applied, the LED and buzzer should be on. (If not, cover or darken the phototransistor.)

18. The phototransistor is most sensitive when viewed from the top. Point the beam of a flashlight directly at the top of the phototransistor. The LED and buzzer should become inactive.

19. Move the flashlight away from and toward the phototransistor a few times to experience the effect that strong light has on the phototransistor.

COMPUTER (OPTIONAL)

20. Repeat Steps 1 to 15 using EWB or an equivalent circuit simulator. Do not record any new values. But make sure that you get reasonable agreement between the EWB measurements and the values recorded earlier.

21. If you are using the CD-ROM version of this book, click on the Assignments menu and select Chap. 10.

ADDITIONAL WORK (OPTIONAL)

22. Have another student insert one of the following troubles into the circuit of Fig. 25-1: open or short any resistor, open any capacitor, open any connecting wire, short the BE, CB, or CE terminals of either transistor. Use only voltage readings of a DMM or an oscilloscope to troubleshoot.

23. Repeat Step 22 several times until you are confident that you can troubleshoot the circuit for various troubles.

Data for Experiment 25

TABLE 25-1. DC VOLTAGES

	Calculated			Measured		
Stage	V_B	V_E	V_C	V_B	V_E	V_C
1						
2						

TABLE 25-2. AC VOLTAGES: $h_{fe} = $ _____

	Calculated			Measured		
Stage	v_b	v_e	v_c	v_b	v_e	v_c
1						
2						

TABLE 25-3. TROUBLESHOOTING

		DC Voltages			AC Voltages		
Trouble	Stage	V_B	V_E	V_C	v_b	v_e	v_c
Open C_4							
Shorted R_4							
Shorted R_{10}							
Open R_3							
Open C_5							

TABLE 25-4. CRITICAL THINKING: $r_E = $ _____

$A_1 = $

$A_2 = $

$A = $

Questions for Experiment 25

1. The calculated dc base voltage of the first stage was approximately: ( )
   (a) 1.1 V; (b) 1.8 V; (c) 6.28 V; (d) 10 V.
2. The measured dc collector voltage of the second stage was closest to: ( )
   (a) 1.1 V; (b) 1.8 V; (c) 6.28 V; (d) 10 V.
3. The ac base voltage of the first stage was closest to: ( )
   (a) 5 mV; (b) 12 mV; (c) 100 mV; (d) 1.4 V.
4. The ac emitter voltage of the second stage was closest to: ( )
   (a) 5 mV; (b) 12 mV; (c) 100 mV; (d) 1.4 V.
5. The voltage gain from the base of the first stage to the collector of the second stage ( )
   was closest to:
   (a) 10; (b) 115; (c) 230; (d) 1000.

6. Explain why the signal decreased when the coupling capacitor was reconnected in Step 7.

TROUBLESHOOTING

7. Explain what happens when an emitter bypass capacitor opens.

8. Suppose you get a collector-emitter short in the first stage of Fig. 25-1. What is the approximate input impedance looking into the base of the first stage? Explain your answer.

CRITICAL THINKING

9. There is a simple way to modify Fig. 25-1 for use with *pnp* transistors. Explain what changes need to be made.

10. Optional. Instructor's question.

Class A Amplifiers

In a class A amplifier, the transistor operates in the active region throughout the ac cycle. This is equivalent to saying the signal does not drive the transistor into either saturation or cutoff on the ac load line. With a CE amplifier, MPP is the smaller of $2I_{CQ}r_c$ or $2V_{CEQ}$. Some of the important quantities in a class A amplifier are the current drain, the maximum transistor power dissipation, the maximum unclipped load power, and the stage efficiency. In this experiment you will calculate and measure the voltages, currents, and powers of a class A amplifier.

Required Reading

Chapter 11 (Secs. 11-1 to 11-3) of *Electronic Principles,* 6th ed.

Equipment

1   audio generator
1   power supply: 15 V
1   transistor: 2N3904
7   ½-W resistors: 220 Ω, 1 kΩ, 1.5 kΩ, 1.8 kΩ, 2.2 kΩ, 3.9 kΩ, 10 kΩ
3   capacitors: two 1 μF, 470 μF (16-V rating or better)
1   VOM (analog or digital multimeter)
1   oscilloscope

Procedure

CE AMPLIFIER

1. In Fig. 26-1, calculate the quiescent collector current and the quiescent collector-emitter voltage. Record your answers in Table 26-1.
2. Calculate and record the MPP and the current drain of the stage.
3. Calculate the maximum transistor power dissipation, maximum unclipped load power, dc input power to the stage, and stage efficiency. Record your answers in the theoretical column of Table 26-2.
4. Connect the circuit. Reduce the signal generator to zero. Use the VOM to measure I_{CQ} and V_{CEQ}. Record the data.
5. Use the oscilloscope to look at the load voltage. Adjust the signal generator until clipping starts on either

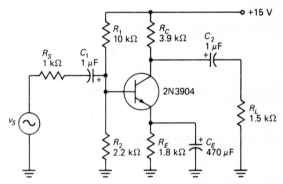

Figure 26-1

half cycle. Notice how the waveform appears squashed on the upper half cycle and elongated on the lower half cycle. This nonlinear distortion is being caused by the large changes in r'_e as the collector approaches cutoff and saturation.

6. Reduce the signal generator until the clipping stops. You have to estimate this as best you can because the clipping is soft as the operating point approaches cutoff. Back off enough from the clipping so that the upper peak appears rounded and smooth. Measure and record the peak-to-peak ac voltage. (This measured value is a rough approximation of the MPP value.)
7. Measure and record the total current drain of the stage.
8. Calculate and record the experimental quantities listed in Table 26-2 using the measured data of Table 26-1.
9. Adjust the signal generator to get a peak-to-peak load voltage of 2 V. Notice how much nonlinear distortion there is.
10. Insert a swamping resistor of 220 Ω. Adjust the signal generator to get a load voltage of 2 V pp and notice how

the load signal appears less distorted than before. You should know from your textbook why this improvement occurs.

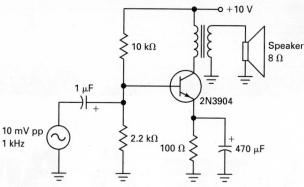

Figure 26-2

TROUBLESHOOTING

11. In Fig. 26-1, assume that R_2 is shorted. Calculate the MPP value and current drain for this trouble. Record in Table 26-3.
12. Repeat Step 11 for each trouble listed in Table 27-3.
13. Connect the circuit with each trouble. Measure and record MPP and I_S.

CRITICAL THINKING

14. Select a value of R_E to get maximum MPP value in Fig. 26-1. Record the nearest standard resistance at the top of Table 26-4. Calculate and record the other quantities of Table 26-4.
15. Connect the circuit with your design value of R_E. Measure and record MPP and I_S in Table 26-4. Calculate and record the experimental quantities $P_{L(max)}$, P_S, and η using the measured data for MPP and I_S.

APPLICATION (OPTIONAL)

16. Connect the circuit shown in Fig. 26-2 using a small 8-Ω speaker connected across the secondary of an audio transformer. The audio transformer should be rated as 600-Ω primary impedance and 8-Ω secondary impedance.
17. Adjust the frequency of the audio generator to 1 kHz. Adjust the output level of the audio generator to produce an ac base voltage of 10 mV pp. You should hear an output from the speaker. Vary the frequency and notice that the tone changes.
18. Measure the peak-to-peak collector voltage. It should be from 0.5 to 2 V pp. Calculate the voltage gain.

19. Measure the peak-to-peak voltage across the speaker. Calculate the speaker power.

COMPUTER (OPTIONAL)

20. Repeat Steps 1 to 15 using EWB or an equivalent circuit simulator. Do not record any new values. But make sure that you get reasonable agreement between the EWB measurements and the values recorded earlier.
21. If you are using the CD-ROM version of this book, click on the Assignments menu and select Chap. 11.

ADDITIONAL WORK (OPTIONAL)

22. Have another student insert one of the following troubles into the circuit of Fig. 26-1: open or short any resistor, open any capacitor, open any connecting wire, short the BE, CB, or CE terminals of either transistor. Use only voltage readings of a DMM or an oscilloscope to troubleshoot.
23. Repeat Step 22 several times until you are confident that you can troubleshoot the circuit for various troubles.

Data for Experiment 26

TABLE 26-1. CE AMPLIFIER

	Calculated	Measured
I_{CQ}		
V_{CEQ}		
MPP		
I_S		

TABLE 26-2. POWER AND EFFICIENCY

	Theoretical	Experimental
$P_{D(max)}$		
$P_{L(max)}$		
P_S		
η		

TABLE 26-3. TROUBLESHOOTING

Trouble	Estimated MPP	I_S	Measured MPP	I_S
Shorted R_2				
Open C_E				
Open R_L				
Collector-emitter open				

TABLE 26-4. CRITICAL THINKING: $R_E =$ _____

	Theoretical	Experimental
MPP		
I_S		
$P_{L(max)}$		
P_S		
η		

Questions for Experiment 26

1. The theoretical MPP value of Fig. 26-1 is approximately:　　　　　　( )
　　(**a**) 1.1V;　　(**b**) 2.35 V;　　(**c**) 9 V;　　(**d**) 15 V.

2. The total current drain of the amplifier was closest to:　　　　　( )
　　(**a**) 1.1 mA;　　(**b**) 2.3 mA;　　(**c**) 4.8 mA;　　(**d**) 6.9 mA.

3. The maximum transistor power dissipation of Fig. 26-1 is approximately:　( )
　　(**a**) 0.46 mW;　　(**b**) 10 mW;　　(**c**) 35.1 mW;　　(**d**) 50 mW.

4. Theoretically, the maximum efficiency of Fig. 26-1 is approximately: ( )
  (a) 0; **(b)** 1.3%; **(c)** 5%; **(d)** 25%.

5. Inserting a swamping resistor in the circuit of Fig. 26-1: ( )
  (a) reduced supply voltage; **(b)** increased quiescent collector current;
  (c) decreased nonlinear distortion; **(d)** increased ac output compliance.

6. Explain why nonlinear distortion exists in a CE amplifier with a large output signal.

TROUBLESHOOTING

7. What happens to the MPP value when the bypass capacitor CE opens? To the voltage gain?

8. Why did the MPP value increase when R_L was opened?

CRITICAL THINKING

9. Explain how you selected the value of R_E to get maximum MPP.

10. Optional. Instructor's question.

Class C Amplifiers

In a class C amplifier, the transistor operates in the active region for less than 180° of the ac cycle. Typically, the conduction angle is much smaller than 180° and the collector current is a train of narrow pulses. This highly nonsinusoidal current contains a fundamental frequency plus harmonics. A tuned class C amplifier has a resonant tank circuit that is tuned to the fundamental frequency. This produces a sinusoidal output voltage of frequency f_r. In a frequency multiplier, the resonant tank circuit is tuned to the nth harmonic, so that the sinusoidal output has a frequency of nf_r.

In this experiment you will build a tuned class C amplifier and a frequency multiplier.

Required Reading

Chapter 11 (Secs. 11-5 and 11-6) of *Electronic Principles*, 6th ed.

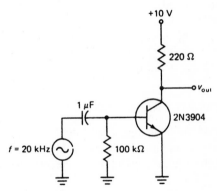

Figure 27-1

Equipment

1   RF generator
1   power supply: 10 V
1   transistor: 2N3904
3   ½-W resistors: 220 Ω, two 100 kΩ
1   inductor: 33 mH (or smaller value as close as possible)
3   capacitors: two 1 μF, 470 pF
1   VOM (analog or digital multimeter)
1   oscilloscope
1   frequency counter (optional)

Procedure

1. Connect the circuit of Fig. 27-1.
2. Set the input frequency of 90 kHz. Adjust the signal level to get narrow output pulses with a peak-to-peak value of 6 V.
3. Measure the pulse width W and period T. Record these values in Table 27-1. Calculate and record the duty cycle.
4. Look at the signal on the base. You should see a negatively clamped waveform. With the oscilloscope on dc input, measure and record the positive and negative peak voltages of this clamped signal.

5. In Fig. 27-2, calculate the resonant frequency of the tuned amplifier. Record your answer in Table 27-2.
6. Assume that the Q_L of the coil is 15. Calculate and record the other quantities listed in Table 27-2.
7. Connect the circuit. Use the oscilloscope to look at the base signal. Adjust the frequency to 40 kHz and the signal level to 2 V pp.
8. Use the oscilloscope to look at the collector signal. Vary the input frequency until the output signal reaches a maximum value (resonance).
9. Adjust the signal level as needed to get 15 V pp at the collector.
10. Repeat Steps 8 and 9 until the circuit is resonant with an output of 15 V pp. (This repetitive adjustment of frequency and signal level is called "rocking in.")

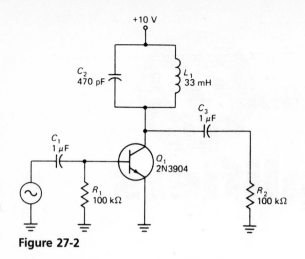

Figure 27-2

RESONANT FREQUENCY, BANDWIDTH, AND CIRCUIT Q

11. Use the frequency counter to measure the resonant frequency. If a frequency counter is not available, use the oscilloscope or generator indication. Record in Table 27-3.

12. Measure f_1 and f_2. Calculate and record the bandwidth in Table 27-3.

13. Readjust the input frequency to get resonance. Calculate the circuit Q using the f_r and B of Table 27-3. Record the circuit Q.

MPP, CURRENT DRAIN, AND DC CLAMPING

14. Increase the input signal level until the output signal just starts clipping. Back off slightly from this level until the signal stops clipping. Record the MPP in Table 27-3.

15. Use the VOM as an ammeter to measure the current drain of the circuit. Record I_S.

FREQUENCY MULTIPLIER

16. Reduce the input frequency and notice that the output signal decreases (off resonance). Continue decreasing the input frequency until the output signal again reaches a maximum value (resonant to a harmonic). Use the frequency counter or oscilloscope to measure the input and output frequencies. Record in Table 27-4. Divide f_{out} by f_{in}, round the answer off to the nearest integer, and notice that it equals 2. The circuit is now operating as a X2 frequency multiplier.

17. Again decrease the input frequency until you find another resonance. Measure and record the input and output frequencies. This time, the f_{out}/f_{in} ratio should be approximately 3. The circuit is now acting like a X3 frequency multiplier.

TROUBLESHOOTING (OPTIONAL)

18. In Fig. 27-2, assume that R_1 is open. Estimate the ac load voltage for this trouble. Record in Table 27-5.

19. Repeat Step 18 for each trouble listed in Table 27-5.

20. Connect the circuit with each trouble. Measure and record the ac load voltage.

CRITICAL THINKING

21. Select a value of C_2 (nearest standard value) to get a resonant frequency of approximately 250 kHz.

22. Connect the circuit with your design value of C_2. Tune to the fundamental frequency. Record the capacitance and frequency here:

$C_2 = $ _____ .

$f_r = $ _____ .

COMPUTER (OPTIONAL)

23. Repeat Steps 1 to 22 using EWB or an equivalent circuit simulator. Do not record any new values. But make sure that you get reasonable agreement between the EWB measurements and the values recorded earlier.

24. If you are using the CD-ROM version of this book, click on the Assignments menu and select Chap. 11.

ADDITIONAL WORK (OPTIONAL)

25. When testing RF circuits with an oscilloscope, you must be constantly aware of the loading effect of the probes. An inexpensive oscilloscope may add as much as 200 pF to a circuit on the X1 probe position. If you use the X10 probe position, the input capacitance decreases to 20 pF. Therefore, use the X10 position whenever possible.

26. Probe capacitances will depend on the quality of the oscilloscope you are using. Find out what the probe capacitances are for the X1 and X10 position of the oscilloscope you are using. Record the values:

$C_{(X1)} = $ _____ .

$C_{(X10)} = $ _____ .

27. Replace the 33-mH inductor by a 330-μH inductor (or an inductor near this value). Since the resonant frequency is inversely proportional to the square root of inductance, the new resonant frequency will be approximately 400 kHz.

28. Look at the output with the X1 position of the probe. Measure the resonant frequency and record the value:

$f_r = $ _____ (X1)

29. Switch the probe to the X10 position. Increase the sensitivity by a factor of 10 to compensate for the probe attenuation of X10.

30. Repeat Step 28 and record the resonant frequency:

$f_r =$ _____ (X10)

31. The frequency in Step 30 should be slightly higher than the frequency in Step 28 because the probe will be adding less capacitance to the circuit during the measurement of resonant frequency.

32. When no probe is connected, is the resonant frequency higher, lower, or the same as the frequency in Step 30?

Answer = _____ .

33. What have you learned about the effect that an oscilloscope probe has on a circuit under test conditions?

34. If you were to shorten all the wires as much as possible in your circuit, what effect would this have on the resonant frequency? Why do you say this?

Data for Experiment 27

TABLE 27-1. WAVEFORMS

$W =$

$T =$

$D =$

$+$ peak $=$

$-$ peak $=$

TABLE 27-2. CALCULATIONS FOR TUNED AMPLIFIER

$f_r =$

$X_L =$

$R_S =$

$R_P =$

$r_C =$

$Q =$

$B =$

$MPP =$

$P_{L(\max)} =$

TABLE 27-3. MEASUREMENTS FOR TUNED AMPLIFIER

$f_r =$

$B =$

$Q =$

$MPP =$

$I_S =$

TABLE 27-4. FREQUENCY MULTIPLIER

f_{in}	f_{out}	n
		2
		3

TABLE 27-5. TROUBLESHOOTING

Trouble	Estimated v_{out}	Measured v_{out}
R_1 open		
Q_1 collector-emitter short		
C_2 short		
C_3 open		

Questions for Experiment 27

1. The duty cycle of Table 27-1 is closest to: ( )
   (a) 1%;   (b) 10%;   (c) 31.6%;   (d) 75%.
2. The MPP of Table 27-2 is approximately: ( )
   (a) 0.7 V;   (b) 1.4 V;   (c) 10 V;   (d) 20 V.
3. To calculate the total dc input power to the tuned amplifier, we can multiply the ( )
   measured current drain of Table 27-3 by the:
   (a) MPP;   (b) circuit Q;   (c) supply voltage;   (d) bandwidth.
4. Assume the current drain equals 0.25 mA in Fig. 27-2. If the ac output compliance ( )
   is 19.6 V, then the stage efficiency is approximately:
   (a) 5%;   (b) 19%;   (c) 47%;   (d) 73%.
5. When f_{in} = 4.75 kHz and f_{out} = 19 kHz, a frequency multiplier is tuned to which ( )
   harmonic of the fundamental frequency?
   (a) first;   (b) second;   (c) third;   (d) fourth.
6. Briefly explain how a tuned class C amplifier works.

7. Explain how a frequency multiplier works.

TROUBLESHOOTING

8. The input voltage driving the tuned class C amplifier of Fig. 27-2 is 1 V pp. The output
   voltage is zero. Name the most likely trouble.

CRITICAL THINKING

9. In this experiment, the tuned class C amplifier has a stage efficiency of only 20 percent,
   more or less depending on the components used. What do you think was the cause of this
   poor efficiency? Explain your answer.

10. Optional. Instructor's question.

The Emitter Follower

An emitter follower has high input impedance, low output impedance, and low non-linear distortion. It is often used as a buffer stage between a high-impedance source and a low-resistance load. In this experiment you will build an emitter follower to verify its high input impedance and low output impedance.

Required Reading

Chapter 12 (Secs. 12-1 to 12-3) of *Electronic Principles,* 6th ed.

Equipment

1 audio generator
1 power supply: 10 V
1 transistor: 2N3904 (or equivalent)
5 ½-W resistors: 51 Ω, 3.9 kΩ, 4.7 kΩ, two 10 kΩ
2 capacitors: 1 μF, 470 μF (10-V rating or better)
1 VOM (analog or digital multimeter)
1 oscilloscope

Procedure

EMITTER FOLLOWER

1. In Fig. 28-1, calculate the dc voltage at the base, emitter, and collector. Record your answers in Table 28-1.
2. Refer to the 2N3904 data sheet to determine the typical h_{fe}. Record this value in Table 28-1.
3. Calculate and record the peak-to-peak ac voltage at the base, emitter, and collector.

4. Connect the circuit. Measure and record the dc voltage at the base, emitter, and collector.
5. Adjust the signal generator to get a source signal of 1 V peak-to-peak at 10 kHz. (Measure this between the left end of the source resistor and ground.)
6. Measure and record the peak-to-peak ac voltage at the base, emitter, and collector.

OUTPUT IMPEDANCE

7. Calculate the theoretical output impedance for the circuit of Fig. 28-1. Record your answer in Table 28-2.
8. Reduce the peak-to-peak source from 1 V to 100 mV.
9. Measure and record the peak-to-peak output voltage (unloaded).
10. Connect a load resistance of 51 Ω across the output.
11. Measure the loaded output voltage.
12. Calculate the output impedance of the emitter follower with the data of Steps 9 and 11. Record your experimental answer in Table 28-1.

TROUBLESHOOTING

13. In Fig. 28-1, assume that R_1 is open. Estimate the dc and ac voltages at the output emitter. Record your answers in Table 28-3.
14. Repeat Step 13 for each trouble listed in Table 28-3.
15. Connect the circuit with each trouble. Measure and record the dc and ac voltages.

CRITICAL THINKING

16. In Fig. 28-1, select a value of R_E to get a dc emitter current of 2.5 mA. Record your design value in Table 28-4.

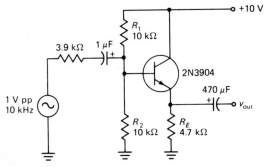

Figure 28-1

117

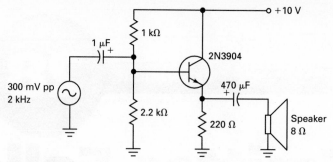

Figure 28-2

17. Connect the circuit with your design value of R_E. Measure and record the dc emitter voltage. Measure and record the dc emitter current.

APPLICATION (OPTIONAL)

18. In Fig. 28-2, the output impedance of the emitter follower is low enough to drive an 8-Ω speaker without using an audio transformer.
19. Connect the circuit shown in Fig. 28-2 using a small 8-Ω speaker. Adjust the frequency of the audio generator to 2 kHz. Adjust the output level of the audio generator to produce an ac base voltage of 300 mV pp. You should hear an output from the speaker. Vary the frequency and notice that the tone changes.
20. Measure the peak-to-peak voltage across the speaker. Calculate the voltage gain of the emitter follower and the speaker power.
21. The Radio Shack 273-223 is a small dc motor that has the following specifications: 1.5 to 3 V dc input with a full-load current of 1 A and a no-load current of 150 to 300 mA. You will be running the motor under no-load conditions since the motor will not be connected to a mechanical load.
22. In Fig. 28-3, the 2N3055 is a power transistor that can produce enough output current for the motor. The four 1N4001s are a diode clamp to protect the motor from excessive input voltage.
23. Connect the circuit of Fig. 28-3. After the circuit is connected, you should be able to see and hear the motor running. If not, adjust the potentiometer until the motor is running.
24. Vary the potentiometer to change the motor speed from off to fully on, and vice versa. Firmly attach a piece of electrical tape to the motor shaft, leaving a ½-inch flap. Now, run the motor at very slow speed. The tape should allow you to see the direction of rotation.

25. Measure the dc voltage across the motor and the dc current through the motor with the motor running at maximum speed. Calculate the motor power.
26. Reverse the motor leads. The motor shaft will turn in the opposite direction. If you vary the motor speed down to zero, you should be able to see the tape turning the opposite way.
27. Questions: What is the voltage across four clamping diodes at maximum speed? What is the power dissipation in the 2N3055? What did you learn?

COMPUTER (OPTIONAL)

28. Repeat Steps 1 to 17 using EWB or an equivalent circuit simulator. Do not record any new values. But make sure that you get reasonable agreement between the EWB measurements and the values recorded earlier.
29. If you are using the CD-ROM version of this book, click on the Assignments menu and select Chap. 12.

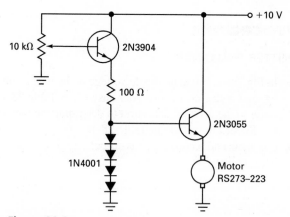

Figure 28-3

Data for Experiment 28

TABLE 28-1. EMITTER FOLLOWER: h_{fe} = _____

	Calculated				Measured		
	B	E	C		B	E	C
dc							
ac							

TABLE 28-2. OUTPUT IMPEDANCE

Calculated r_{out} =	
Unloaded v_{out} =	
Loaded v_{out} =	
Experimental r_{out} =	

TABLE 28-3. TROUBLESHOOTING

	Estimated		Measured	
Trouble	V_E	v_e	V_E	v_e
Open R_1				
Shorted R_1				
Open R_2				
Shorted R_2				
Open R_E				
Shorted R_E				

TABLE 28-4. CRITICAL THINKING

Calculated R_E =	
Measured V_E =	
Measured I_E =	

Questions for Experiment 28

1. The data of Table 28-1 show that the voltage gain of the emitter follower was approximately: ( )
   (a) 0;   (b) 1;   (c) 4.3 V;   (d) 10 V.
2. The ac collector voltage of the emitter follower was closest to: ( )
   (a) 0;   (b) 0.58 V;   (c) 1 V;   (d) 10 V.
3. Because the ac emitter voltage approximately equals the ac base voltage in Table 28-1, the input impedance of the base must be: ( )
   (a) 0;   (b) very low;   (c) 10 V;   (d) very high.
4. The calculated output impedance in Table 28-2 is closest to: ( )
   (a) 1 Ω;   (b) 23 Ω;   (c) 42.4 Ω;   (d) 51 Ω.
5. The unloaded output voltage in Table 28-2 is closest to: ( )
   (a) 0 V;   (b) 10 mV;   (c) 30 mV;   (d) 58 mV.

6. Explain how you worked out the experimental value of r_{out} in Table 28-2.

TROUBLESHOOTING

7. Explain the dc and ac output voltage you got when R_2 was shorted.

8. If there is a collector-emitter short in Fig. 28-1, what happens to the input impedance of the emitter-follower?

CRITICAL THINKING

9. How did you arrive at your design value of R_E?

10. Optional. Instructor's question.

120

Class B Push-Pull Amplifiers

In a class B push-pull amplifier, each transistor operates in the active region for half of the ac cycle. With a single power supply, the MPP is approximately equal to V_{CC}. Class B push-pull amplifiers are widely used for the output stage of audio systems because they can deliver more load power than class A amplifiers. In fact, the theoretical efficiency of a class B push-pull amplifier approaches 78.5 percent, far greater than class A.

The main problem with class B push-pull amplifiers is setting up a stable Q point near cutoff. The required V_{BE} drop for each transistor is in the vicinity of 0.6 to 0.7 V, with the exact value determined by the quiescent collector current needed to avoid crossover distortion. Since collector current increases by a factor of 10 for each V_{BE} increase of 60 mV, setting up a stable and precise I_{CQ} is difficult. Voltage-divider bias is not practical because the Q point is too sensitive to changes in supply voltage, temperature, and transistor replacement. As discussed in the textbook, there is a real danger of thermal runaway. Diode bias is the usual way to bias a class B push-pull amplifier.

In this experiment we will use 1N4148s (or 1N914s) for the compensating diodes. The complementary transistors used are the 2N3904 *(npn)* and 2N3906 *(pnp)*. With the discrete devices of this experiment, the match between the diodes and transistors is only an approximation. For this reason, 10-Ω emitter feedback resistors are used to prevent the excessive collector current that could result from the mismatch between the diodes and the transistors.

Required Reading

Chapter 12 (Secs. 12-5 to 12-7) of *Electronic Principles*, 6th ed.

Equipment

1   audio generator
1   power supply: adjustable from 0 to 15 V
2   diodes: 1N4148 or 1N914
2   transistors: 2N3904, 2N3906
5   ½-W resistors: 100 Ω, two 680 Ω, two 4.7 kΩ
3   capacitors: two 1 μF, 100-μF (16-V rating or better)
2   VOMs (or a dc milliammeter and dc voltmeter)
1   oscilloscope

Procedure

CROSSOVER DISTORTION

1. Adjust your power supply to 5 V and then connect the circuit of Fig. 29-1*a* with this supply voltage.
2. Set the input frequency to 1 kHz and the signal level across the audio generator at 2 V pp.
3. Look at the output signal across the load resistor (100 Ω). You should see crossover distortion.

SINGLE-SUPPLY CIRCUIT

4. Connect the circuit of Fig. 29-2. Reduce the output of the audio generator to 0 V. Measure the current through the diodes. Record this measurement in Table 29-1.

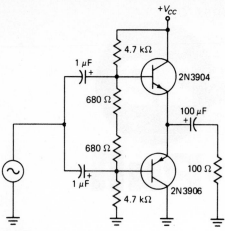

Figure 29-1

7. Measure and record the peak-to-peak output voltage across the 100-Ω load resistor.
8. Increase the output level of the audio generator until clipping just starts on the final output signal. This is the approximate value of the MPP. Record the MPP in Table 29-1.
9. Repeat steps 4 to 8 for two more pairs of transistors.

DUAL-SUPPLY CIRCUIT

10. Connect the circuit of Fig. 29-3. Reduce the output of the audio generator to 0 V. Measure the current through the diodes. Record this measurement in Table 29-2.
11. Measure and record the collector current in the upper 2N3904.
12. Adjust the output of the audio generator to 2 V pp. Record this value in Table 29-2.
13. Measure and record the peak-to-peak output voltage across the 100-Ω load resistor.
14. Increase the output level of the audio generator until clipping just starts on the final output signal. This is the approximate value of the MPP. Record the MPP in Table 29-2.
15. Repeat steps 10 to 14 for two more pairs of transistors.

5. Measure and record the collector current in the upper 2N3904. The diode current and transistor currents of Table 29-1 may differ substantially. The main reason is the mismatch in characteristics between discrete diodes and transistors. With integrated circuits, these two currents would be equal to a close approximation.
6. Adjust the output of the audio generator to 2 V pp. Record this value in Table 29-1.

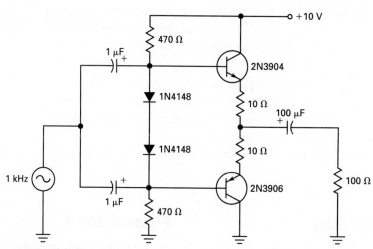

Figure 29-2

122

APPLICATION (OPTIONAL)

16. Connect the circuit shown in Fig. 29-4 using a small 8-Ω speaker. Adjust the frequency of the audio generator to 2 kHz. Adjust the output level of the audio generator to produce an ac input voltage of 100 mV pp. You should hear an output from the speaker. Vary the frequency and notice that the tone changes.

COMPUTER (OPTIONAL)

17. Repeat Steps 1 to 14 using EWB or an equivalent circuit simulator. Do not record any new values. But make sure that you get reasonable agreement between the EWB measurements and the values recorded earlier.

18. If you are using the CD-ROM version of this book, click on the Assignments menu and select Chap. 12.

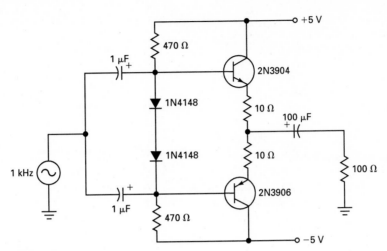

Figure 29-3

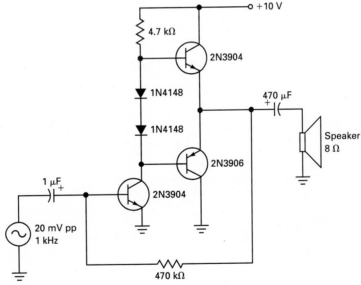

Figure 29-4

Data for Experiment 29

TABLE 29-1. SINGLE-SUPPLY CIRCUIT

Transistors	I_{diode}	I_C	v_{in}	v_{out}	MPP
1					
2					
3					

TABLE 29-2. DUAL-SUPPLY CIRCUIT

Transistors	I_{diode}	I_C	v_{in}	v_{out}	MPP
1					
2					
3					

Questions for Experiment 29

1. In Fig. 29-1, crossover distortion will occur when the supply voltage is: ( )
   (a) negative;   (b) too low;   (c) too high;   (d) unstable.
2. Crossover distortion can be reduced by having: ( )
   (a) low supply voltage;   (b) two *npn* transistors;   (c) a small quiescent I_C;   (d) no load resistor.
3. In Fig. 29-2, the 10-Ω resistors produce: ( )
   (a) negative feedback;   (b) positive feedback;   (c) crossover distortion;   (d) more MPP.
4. In Fig. 29-2, the 1N4148 diodes are: ( )
   (a) germanium;   (b) zeners;   (c) forward-biased;   (d) good matches for the transistors.
5. In Fig. 29-3, the dc voltage between the 10-Ω resistors ideally is: ( )
   (a) 0;   (b) 5 V;   (c) −5 V;   (d) 10 V.
6. In Fig. 29-4, the maximum speaker power is measured in: ( )
   (a) nanowatts;   (b) microwatts;   (c) milliwatts;   (d) watts.
7. Explain why the diode and transistor currents are different in Table 29-1.

8. Explain how the diodes in Fig. 29-2 can help prevent thermal runaway by the transistors.

9. Optional. Instructor's question.

The Zener Follower

B y cascading a zener diode and an emitter follower, we can build voltage regulators with large load currents. An improved regulator like this can hold the load voltage almost constant despite large changes in load current because the circuit appears stiff over a greater range of load resistance. The zener follower is an example of a series regulator, one whose load current flows through a pass transistor. Because of their simplicity, series regulators are widely used.

In this experiment you will build a regulated power supply consisting of a bridge rectifier, a capacitor-input filter, and a zener follower.

Required Reading

Chapter 12 (Sec. 12-8) of *Electronic Principles*, 6th ed.

Equipment

1 center-tapped transformer, 12.6 V ac (Triad F-25X or equivalent) with fused line cord
4 silicon diodes: 1N4001 (or equivalent)
1 zener diode: 1N757 (or equivalent 9-V zener diode)
1 transistor: 2N3055 (or equivalent power transistor)
2 ½-W resistors: 1 kΩ, 10 kΩ
2 1-W resistors: 100 Ω, 820 Ω
1 capacitor: 470 μF (25-V rating or better)
1 VOM (analog or digital multimeter)
1 oscilloscope

Procedure

REGULATED POWER SUPPLY

1. A 1N757 has a nominal zener voltage of 9.1 V. In Fig. 30-1, calculate the input voltage, zener voltage, and output voltage for the zener follower. (The input voltage is across the filter capacitor.) Record your answers in Table 30-1.
2. Connect the regulated power supply of Fig. 30-1.
3. Measure and record all voltages listed in Table 30-1.

VOLTAGE REGULATION

4. Estimate and record the output voltages in Fig. 30-1 for each of the load resistors listed in Table 30-2.

5. Measure and record the output voltages for the load resistances of Table 30-2.

RIPPLE ATTENUATION

6. For each load resistance listed in Table 30-3, calculate and record the peak-to-peak ripple across the filter capacitor. Also calculate and record the peak-to-peak ripple at the output. (Assume a zener resistance of 10 Ω.)
7. For each load resistance of Table 30-3, measure and record the peak-to-peak ripple at the input and output of the zener follower. (*Note:* If the ripple appears very fuzzy or erratic, you may have oscillations. Try shortening the leads. If that doesn't work, consult the instructor.)

TROUBLESHOOTING

8. Assume that D_1 is open.
9. Estimate the input and output voltage of the zener follower for the foregoing trouble. Record your answers in Table 30-4.
10. Connect the circuit with the foregoing trouble. Measure and record the input and output voltage.
11. Repeat Steps 9 and 10 for the other troubles listed in Table 30-4.

CRITICAL THINKING

12. Modify the power supply so that it produces a regulated output voltage of approximately 5.5 V for a load current between 0 and 55 mA. The zener current should be approximately 20 mA. Record your design changes in Table 30-5.

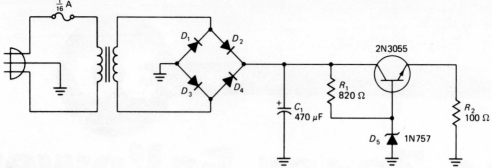

Figure 30-1

13. Calculate and record the input voltage, input ripple, output voltage, and output ripple for the zener follower with a load resistance of 100 Ω. Assume the zener diode has a zener resistance of 7 Ω.

14. Connect your design with a load resistance of 100 Ω. Measure all dc voltages and ripples listed in Table 30-5. Record your data.

COMPUTER (OPTIONAL)

15. Repeat Steps 1 to 14 using EWB or an equivalent circuit simulator. Do not record any new values. But make sure that you get reasonable agreement between the EWB measurements and the values recorded earlier.

16. If you are using the CD-ROM version of this book, click on the Assignments menu and select Chap. 12.

ADDITIONAL WORK (OPTIONAL)

17. Have another student insert one of the following troubles into the circuit: open any component or connecting wire. Use only voltage readings of a DMM or an oscilloscope to troubleshoot.

18. Repeat Step 17 several times until you are confident that you can troubleshoot the circuit for various troubles.

19. Measure the load voltage and load current for several values of load resistance between 100 Ω and 10 kΩ. Record your data on a separate piece of paper.

20. With the data of Step 19, graph V_L versus I_L on semilog paper.

21. With the data of Step 19, graph V_L versus R_L on semilog paper.

Data for Experiment 30

TABLE 30-1. REGULATED POWER SUPPLY

	V_{in}	V_Z	V_{out}
Calculated			
Measured			

TABLE 30-2. VOLTAGE REGULATION

R_L	Estimated V_{out}	Measured V_{out}
100 Ω		
1 kΩ		
10 kΩ		

TABLE 30-3. RIPPLE

	Calculated V_{rip}		Measured V_{rip}	
R_L	In	Out	In	Out
100 Ω				
1 kΩ				
10 kΩ				

TABLE 30-4. TROUBLESHOOTING

	Estimated		Measured	
	V_{in}	V_{out}	V_{in}	V_{out}
Open D_1				
Open R_1				
Shorted D_5				
Collector-emitter open				

TABLE 30-5. CRITICAL THINKING

Describe your changes here:

	V_{in}	Input V_{rip}	V_{out}	Output V_{rip}
Calculated				
Measured				

Questions for Experiment 30

1. When the load resistance increases in Table 30-2, the measured output voltage: ( )
   (a) decreases slightly: (b) stays the same; (c) increases slightly;
   (d) none of the foregoing.

2. When the load resistance increases, the input ripple to the zener follower: ( )
   (a) decreases; (b) stays the same; (c) increases; (d) none of
   the foregoing.

3. The pass transistor of Fig. 30-1 has a power dissipation that is closest to: ( )
   (a) 0.25 W; (b) 0.5 W; (c) 0.7 W; (d) 1 W.

4. The zener diode of Fig. 30-1 has a zener current of approximately: ( )
   (a) 1 mA; (b) 2.35 mA; (c) 12.2 mA; (d) 18.5 mA.

5. In Fig. 30-1, the load current is approximately; ( )
   (a) 1 mA; (b) 18.5 mA; (c) 84 mA; (d) 523 mA.

6. Explain how the regulated power supply of Fig. 30-1 works.

TROUBLESHOOTING

7. Explain why the regulator continues to work even though D_1 is opened.

8. Suppose the circuit of Fig. 30-1 has a collector-emitter short (base, emitter, and collector shorted together). What components are likely to be destroyed?

CRITICAL THINKING

9. What changes did you make to meet the requirements of Step 12, and why did you make them?

10. Optional. Instructor's question.

An Audio Amplifier

In this experiment you will build a discrete audio amplifier with a class A input stage and a class B push-pull emitter follower as shown in Fig. 31-1. Adjustment R_7 is included to allow you to set the Q_3 emitter voltage at +5 V, half the supply voltage. This lets the output signal swing equally in both directions to get maximum MPP value. Some of the resistance values have not been optimized to allow you to improve the design (optional). The circuit is fairly complicated, so take your time in wiring it. Double-check your connections before applying power.

Required Reading

Chapters 10 to 12 of *Electronic Principles*, 6th ed.

Equipment

1   audio generator
1   power supply: 10 V
2   diodes: 1N4148 or 1N914
4   transistors: three 2N3904, 2N3906
9   ½-W resistors: two 100 Ω, two 1 kΩ, 2.2 kΩ, 3.9 kΩ, 4.7 kΩ, two 10 kΩ
1   potentiometer: 5 kΩ
4   capacitors: two 1 μF, 100 μF, 470 μF (16-V rating or better)
1   VOM (analog or digital multimeter)
1   oscilloscope

Procedure

AUDIO AMPLIFIER

1. Assume that the base resistance of Q_2 is adjusted to produce a quiescent voltage of +5 V at the emitter of Q_3. Calculate and record all dc voltages listed in Table 31-1.
2. Assume an ac load voltage of 4 V pp. Also assume all h_{fe} values are 100. Calculate and record all ac voltages listed in Table 31-2.
3. Build the audio amplifier of Fig. 31-1. Adjust the base resistor of Q_2 to produce a dc voltage +5 V at the emitter of Q_3.
4. Set the input frequency to 1 kHz and adjust the ac load voltage to 4 V pp.
5. Use the VOM to measure all dc voltages listed in Table 31-1. Use the oscilloscope to measure all ac voltages in Table 31-2. Record your data.

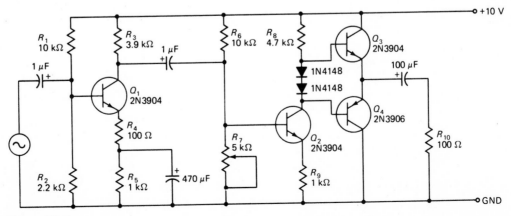

Figure 31-1

TROUBLESHOOTING

6. Ask another student to insert any of the following troubles into the circuit: open any component or connecting wire.
7. Locate and repair the trouble.
8. Repeat Steps 6 and 7 two or more times as desired.

CRITICAL THINKING

9. Measure the MPP value. Record it as the initial value in Table 31-3.
10. Try to figure out how to increase the MPP value. For instance, changing certain resistances will increase the MPP value.
11. Insert your changes and measure the MPP value.
12. When you have increased the MPP value as much as possible, record the final MPP in Table 31-3. Also record the changes you inserted.

APPLICATION (OPTIONAL)

13. Replace the 100-Ω load by an 8-Ω speaker. Adjust the output level of the audio generator to produce an audible tone. Vary the frequency and notice that the tone changes.

COMPUTER (OPTIONAL)

14. Repeat Steps 1 to 5 using EWB or an equivalent circuit simulator. Do not record any new values. But make sure that you get reasonable agreement between the EWB measurements and the values recorded earlier.
15. If you are using the CD-ROM version of this book, click on the Assignments menu and select Chap. 12.

ADDITIONAL WORK (OPTIONAL)

16. Write an essay on how to troubleshoot a circuit like Fig. 31-1. Discuss at least three different troubles and the methods that can be used to isolate the trouble.

Data for Experiment 31

TABLE 31-1. DC VOLTAGES

	Calculated			Measured		
	B	*E*	*C*	*B*	*E*	*C*
Q_1						
Q_2						
Q_3						
Q_4						

TABLE 31-2. AC VOLTAGES

	Calculated			Measured		
	B	*E*	*C*	*B*	*E*	*C*
Q_1						
Q_2						
Q_3						
Q_4						

TABLE 31-3. CRITICAL THINKING

Initial MPP =

Final MPP =

Changes were as follows:

Questions for Experiment 31

1. Resistor R_7 is adjusted to get approximately: ( )
   (a) 10 mA through R_8;   (b) +10 V at the collector of Q_3;   (c) +5 V at the
   emitter of Q_3;   (d) 0 V at the collector of Q_4.
2. The capacitive reactance of 100 μF at 1 kHz is approximately: ( )
   (a) 1.59 Ω;   (b) 6.28 Ω;   (c) 100 Ω;   (d) 1 kΩ.
3. The amplifier of Fig. 31-1 had a voltage gain closest to: ( )
   (a) 1;   (b) 25;   (c) 100;   (d) 200.
4. The measured ac voltage at the base of Q_3 was slightly higher than the ac output   ( )
   voltage because:
   (a) the output capacitor was too small;   (b) of the drop across r'_e;
   (c) R_7 was adjustable;   (d) Q_2 was an *npn* transistor.
5. One reason the MPP value of the output stage is less than V_{CC} is because of: ( )
   (a) voltage drop across X_C;   (b) offset voltage of the 1N914s;   (c) V_{BE}
   drops of the output transistors;   (d) power dissipation by the load resistor.

6. Explain how the audio amplifier of Fig. 31-1 works.

TROUBLESHOOTING

7. Suppose a 1N4148 shorts in Fig. 31-1. What kind of symptoms will this produce?

8. Resistor R_4 of Fig. 31-1 is shorted by a solder bridge. Describe some of the symptoms that result.

CRITICAL THINKING

9. Why is the MPP value less than V_{CC} in Fig. 31-1?

10. Why did the changes that you made to Fig. 31-1 increase the MPP?

134

Experiment

JFET Bias

$\mathbf{G}$ate bias is the simplest but worst way to bias a JFET for linear operation because the drain current depends on the exact value of V_{GS}. Since V_{GS} has a large variation, the drain current has a large variation. Self-bias offers some improvement because the source resistor produces local feedback which reduces the effect of V_{GS}. When large supply voltages are available, voltage-divider bias results in a relatively stable Q point. Finally, current-source bias can produce the most stable Q point because a bipolar current source sets up the drain current through the JFET.

In this experiment, you will bias a JFET in the different methods just described. This will illustrate the stability of each type of bias.

Required Reading

Chapter 13 (Secs. 13-1 to 13-5) of *Electronic Principles*, 6th ed.

Equipment

2 power supplies: adjustable from 0 to ± 15 V
1 transistor: 2N3904
3 JFETs: MPF102 (or equivalent)
7 ½-W resistors: 470 Ω, 680 Ω, 1 kΩ, 2.2 kΩ, 6.8 kΩ, 33 kΩ, 100 kΩ
1 VOM (analog or digital multimeter)

Procedure

MEASURING I_{DSS}

1. Refer to the data sheet of an MPF102 in the Appendix. Notice that $V_{GS(\text{off})}$ has a maximum of -8 V; the minimum value is not specified. Also notice that I_{DSS} has a minimum value of 2 mA and a maximum value of 20 mA.
2. Connect the circuit of Fig. 32-1a. Measure the drain current. Record this value of I_{DSS} in Table 32-1. (*Note:* Because of heating effects, the drain current may decrease slowly. Take your reading as soon after power-up as possible.)
3. Repeat Step 2 for the other JFETs.

GATE BIAS

4. With gate bias, you apply a fixed gate voltage that reverse-biases the gate of the JFET. This produces a

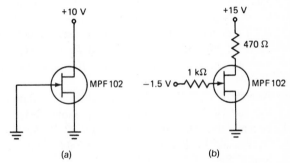

Figure 32-1

drain current that is less than I_{DSS}. The problem is that you cannot accurately predict the drain current in mass production because of the variation in the required V_{GS}. The following steps will illustrate this point.

5. Connect the circuit of Fig. 32-1b. Measure V_{GS}, I_D, and V_D. Record the data in Table 32-2.
6. Repeat Step 5 for the other JFETs. If you have a random set of three JFETs, the drain current usually will show a significant variation from one JFET to another.

MEASURING $V_{GS(\text{OFF})}$

7. Here is an approximate way to measure $V_{GS(\text{off})}$. Insert the first JFET into the gate-biased circuit. Increase the negative gate supply voltage of Fig. 32-1b until the drain current is approximately 1 μA. (If your VOM cannot measure down to 1 μA, then use 10 μA or 100 μA, or whatever low value your instructor indicates.) Record the approximate $V_{GS(\text{off})}$ in Table 32-1.
8. Repeat Step 7 for the other two JFETs.

135

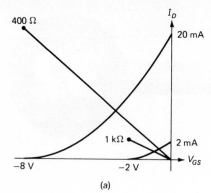

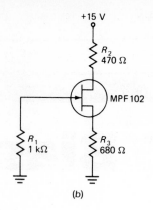

(a)   (b)

Figure 32-2

SELF-BIAS

9. The data sheet of a JFET lists a maximum I_{DSS} of 20 mA and a maximum $V_{GS(off)}$ of -8 V. If a JFET has these values, its transconductance curve appears as shown in Fig. 32-2a. Notice that the approximate source resistance for self-bias is

$$R_S = \frac{8\ V}{20\ mA} = 400\ \Omega$$

The minimum I_{DSS} is 2 mA. For this experiment, we will assume that the minimum $V_{GS(off)}$ is -2 V. A JFET with these values has the lower transconductance curve of Fig. 32-2a, and required R_S is 1 kΩ. An average source resistance is around 700 Ω, so we will use 680 Ω in our test circuit.

10. Assume a V_{GS} of -2 V in Fig. 32-2b. Calculate I_D, V_D, and V_S. Record your answers in Table 32-3.
11. Connect the self-bias circuit of Fig. 32-2b. Measure I_D, V_D, and V_S. Record your data in Table 32-3.
12. Repeat Step 11 for the other JFETs.
13. Notice that the drain current of the self-bias circuit (Table 32-3) has less variation than the drain current of the gate-biased circuit (Table 32-2).

VOLTAGE-DIVIDER BIAS

14. Assume that V_{GS} is -2 V in Fig. 32-3. Calculate I_D, V_D, and V_S. Record your answers in Table 32-4.
15. Connect the circuit. Measure and record the quantities of Table 32-4.

16. Repeat Step 15 for the other JFETs. Notice that the drain current of Table 33-4 has less variation than the drain currents of Tables 32-2 and 32-3.

CURRENT-SOURCE BIAS

17. Assume V_{GS} is -2 V in Fig. 32-4. Calculate and record the quantities of Table 32-5.
18. Connect the circuit. Measure I_D, V_D, and V_S. Record your data.

TROUBLESHOOTING

19. Assume R_2 is shorted in Fig. 32-4. Calculate and record V_D in Table 32-6.
20. Insert the trouble into your circuit. Measure and record V_D.
21. Repeat Steps 19 and 20 for the other troubles listed in Table 32-6.

CRITICAL THINKING

22. Use Eq. (13-1) of your textbook and the data of Table 32-1 to calculate the source resistance of a self-biased circuit. Average the three source resistances. Record your answer at the top of Table 32-7.
23. Connect the circuit of Fig. 32-2b with your design value of R_S. Measure I_D, V_D, and V_S. Record data.
24. Repeat Step 23 for the other JFETs.

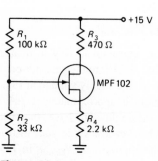

Figure 32-3

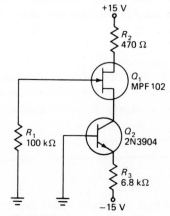

Figure 32-4

136

COMPUTER (OPTIONAL)

25. Repeat Steps 1 to 23 using EWB or an equivalent circuit simulator. Do not record any new values. But make sure that you get reasonable agreement between the EWB measurements and the values recorded earlier.

26. If you are using the CD-ROM version of this book, click on the Assignments menu and select Chap. 13.

ADDITIONAL WORK (OPTIONAL)

27. In this section, you will measure V_{GS}, V_{DS}, and I_D. Then, you will graph the drain curves.

28. In Fig. 32-1a, measure and record I_D for different values of V_{DS}.

29. Remove the short on the gate. Apply a V_{GS} of -1 V and repeat Step 28.

30. Apply several more negative values of V_{GS} and repeat Step 28 until you have enough data for a graph like Fig. 13-5 in your textbook.

31. Graph I_D versus V_{DS} for different values of V_{GS}.

Data for Experiment 32

TABLE 32-1. JFET DATA

JFET	I_{DSS}	$V_{GS(off)}$
1		
2		
3		

TABLE 32-2. GATE BIAS: $V_{GG} = -1.5$ V

JFET	V_{GS}	I_D	V_D
1			
2			
3			

TABLE 32-3. SELF-BIAS

JFET	Calculated I_D	Calculated V_D	Calculated V_S	Measured I_D	Measured V_D	Measured V_S
1						
2						
3						

TABLE 32-4. VOLTAGE-DIVIDER BIAS

JFET	Calculated I_D	Calculated V_D	Calculated V_S	Measured I_D	Measured V_D	Measured V_S
1						
2						
3						

TABLE 32-5. CURRENT-SOURCE BIAS

JFET	Calculated I_D	Calculated V_D	Calculated V_S	Measured I_D	Measured V_D	Measured V_S
1						
2						
3						

TABLE 32-6. TROUBLESHOOTING

Trouble	Estimated		Measured	
	V_D	V_C	V_D	V_C
R_2 shorted				
Q_2 collector-emitter short				
R_3 open				

TABLE 32-7. DESIGN: $R_S =$ _____

JFET	I_S	V_D	V_S
1			
2			
3			

Questions for Experiment 32

1. This experiment proved that the circuit with the least variation in drain current was:　( )
   (a) gate bias;　(b) self-bias;　(c) voltage-feedback;　(d) current-source bias.
2. With gate bias, the drain current has which of the following:　( )
   (a) almost constant value;　(b) constant drain voltage;　(c) values greater than I_{DSS};　(d) large variations.
3. Self-bias is better than:　( )
   (a) gate bias;　(b) voltage-divider bias;　(c) current-source bias;
   (d) emitter bias.
4. The voltage across the source resistor of Fig. 32-3 equals the gate voltage plus the　( )
   magnitude of:
   (a) V_{GS};　(b) V_D;　(c) V_S;　(d) V_{DD}.
5. The voltage-divider bias of Fig. 32-3 would be more stable if we:　( )
   (a) decreased the supply voltage;　(b) increased the supply voltage;
   (c) decreased R_2;　(d) increased R_1.
6. Briefly discuss the variations in drain current for the four types of JFET bias.

TROUBLESHOOTING

7. Suppose the voltage across the source resistor of Fig. 32-3 is zero. Name three possible troubles.

8. The drain voltage of Fig. 32-2 is 15 V. Name three possible troubles.

CRITICAL THINKING

9. What value of R_S did you use in your design? How did you arrive at this value?

10. Optional. Instructor's question.

JFET Amplifiers

Because the transconductance curve of a JFET is parabolic, large-signal operation of a CS amplifier produces square-law distortion. This is why a CS amplifier is usually operated small-signal. JFET amplifiers cannot compete with bipolar amplifiers when it comes to voltage gain. Because g_m is relatively low, the typical CS amplifier has a relatively low voltage gain.

The CD amplifier, better known as the source follower, is analogous to the emitter follower. The voltage gain approaches unity and the input impedance approaches infinity, limited only by the external biasing resistors connected to the gate. The source follower is a popular circuit often found near the front end of measuring instruments.

In this experiment, you will build a CS amplifier and a source follower to verify the relations discussed in your textbook.

Required Reading

Chapter 13 (Secs. 13-6 and 13-7) of *Electronic Principles*, 6th ed.

Equipment

1 audio generator
1 power supply: 15 V
3 JFETs: MPF102 (or equivalent)
4 ½-W resistors: 1 kΩ, two 2.2 kΩ, 220 kΩ
3 capacitors: two 1 μF, 100 μF (16-V rating or better)
1 potentiometer: 5 kΩ
1 oscilloscope

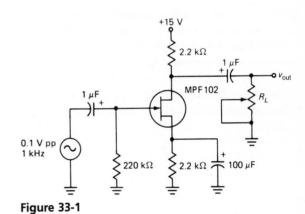

Figure 33-1

Procedure

CS AMPLIFIER

1. Assume that the JFET of Fig. 33-1 has a typical g_m of 2000 μS. Calculate the unloaded voltage gain, output voltage, and output impedance. Record your answers in Table 33-1.
2. Connect the circuit with R_L equal to infinity (no load resistor).
3. Adjust the audio generator to 1 kHz. Set the signal level to 0.1 V pp across the input.
4. Look at the output signal. It should be an amplified sine wave. Measure and record the peak-to-peak output voltage. Then calculate the voltage gain. Record your answer as the measured A in Table 33-1.

5. Connect the 5-kΩ potentiometer as a variable load resistance. Adjust this load resistance until the output voltage is half of the unloaded output voltage.
6. Disconnect the 5-kΩ potentiometer and measure its resistance. Record this as r_{out} in Table 33-1. (*Note:* You have just found the Thevenin or output impedance by the matched-load method.)
7. Repeat Steps 1 to 6 for the other JFETs.

SOURCE FOLLOWER

8. Assume a typical g_m of 2000 μS in Fig. 33-2. Calculate the unloaded voltage gain, output voltage, and output impedance. Record your answers in Table 33-2.

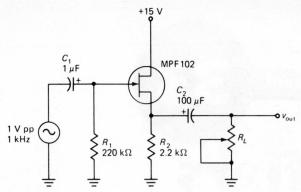

Figure 33-2

9. Connect the circuit with R_L equal to infinity. Adjust the frequency to 1 kHz and the signal level to 1 V pp across the input.
10. Measure and record the output voltage. Calculate the voltage gain and record as the measured A in Table 33-2.
11. Measure and record the output impedance by the matched-load method used earlier.
12. Repeat Steps 8 to 11 for the other JFETs.

TROUBLESHOOTING

13. Table 33-3 lists dc and ac symptoms for Fig. 33-2. Try to figure out what trouble would produce these symptoms. When you think you have the answer, insert the trouble into the circuit. Then measure the dc and ac voltages to verify that your trouble is causing the symptoms. Record the trouble in Table 33-3.
14. Repeat Step 13 for the other symptoms listed in Table 33-3.

CRITICAL THINKING

15. Redesign the source follower of Fig. 33-2 so that it uses voltage-divider bias. Assume V_{GS} is −2 V and select R_1 and R_2 to produce a V_S of +7.5 V. Record your resistance values at the top of Table 33-4.
16. Connect your redesigned source follower. Measure the dc voltage at the source. Record V_S in Table 33-4. Also, set the ac input voltage to 1 V pp. Measure the ac output voltage. Calculate and record the unloaded voltage gain.
17. Repeat Step 16 for each JFET.

COMPUTER (OPTIONAL)

18. Repeat Steps 1 to 16 using EWB or an equivalent circuit simulator. Do not record any new values. But make sure that you get reasonable agreement between the EWB measurements and the values recorded earlier.
19. If you are using the CD-ROM version of this book, click on the Assignments menu and select Chap. 13.

ADDITIONAL WORK (OPTIONAL)

20. What effects do different values of V_{GS} have on the voltage gain and output impedance? This is the question you will answer in this section.
21. Change the source resistor in Fig. 33-1 from 2.2 kΩ to 470 Ω. Measure the voltage gain and output impedance of the circuit.
22. Change the source resistor in Fig. 33-2 from 2.2 kΩ to 470 Ω. Measure the voltage gain and output impedance of the circuit.
23. What did you learn in Steps 21 and 22?

142

Data for Experiment 33

TABLE 33-1. CS AMPLIFIER

	Calculated			Measured		
JFET	v_{out}	A	r_{out}	v_{out}	A	r_{out}
1						
2						
3						

TABLE 33-2. SOURCE FOLLOWER

	Calculated			Measured		
JFET	v_{out}	A	r_{out}	v_{out}	A	r_{out}
1						
2						
3						

TABLE 33-3. TROUBLESHOOTING

DC Symptoms			AC Symptoms				Trouble
V_G	V_D	V_S	v_g	v_d	v_s	v_{out}	
0	15 V	3.7 V	1 V	0	0	0	
0	15 V	3.7 V	1 V	0	0.82 V	0	
0	15 V	3.7 V	0	0	0	0	
0	0	0	1 V	0	0	0	

TABLE 33-4. CRITICAL THINKING: $R_1 =$ _____ ; $R_2 =$ _____

JFET	V_S	A
1		
2		
3		

Questions for Experiment 33

1. The calculated voltage gain of Table 33-1 is approximately: ( )
   (a) 0.44;   **(b)** 1;   **(c)** 4.4;   **(d)** 9.4.
2. The output impedance of Fig. 33-1 is closest to: ( )
   (a) 407 Ω;   **(b)** 2.2 kΩ;   **(c)** 5 kΩ;   **(d)** 220 kΩ.
3. The voltage gain of the source follower was closest to: ( )
   (a) 0.5;   **(b)** 0.8;   **(c)** 1;   **(d)** 4.4.
4. The source follower had an output impedance closest to: ( )
   (a) 0;   **(b)** 100 Ω;   **(c)** 200 Ω;   **(d)** 400 Ω.
5. The main advantage of a JFET amplifier is its: ( )
   (a) high voltage gain;   **(b)** low drain current;   **(c)** high input impedance;
   (d) high transconductance.

6. Compare the voltage gain of a CS amplifier like Fig. 33-1 to a bipolar CE amplifier.

TROUBLESHOOTING

7. In Fig. 33-2, all dc voltages are normal. The ac gate voltage and source voltage are normal. There is no output voltage. What is the most likely trouble?

8. All dc voltages are normal in Fig. 33-2. All ac voltages are zero. What is the most likely trouble?

CRITICAL THINKING

9. How did you arrive at your values of R_1 and R_2?

10. Optional. Instructor's question.

34

JFET Applications

One of the main applications of JFETs is the analog switch. In this application, a JFET acts either like an open switch or like a closed switch. This allows us to build circuits that either transmit an ac signal or block it from the output terminals.

In the ohmic region, a JFET acts like a voltage-variable resistance instead of a current source. This means we can change the value of $r_{ds(on)}$ by changing V_{GS}. When a JFET is used as a voltage-variable resistance, the ac signal should be small, typically less than 100 mV.

Another application of the JFET is with automatic gain control (AGC). Because the g_m of a JFET varies with the Q point, we can build amplifiers whose voltage gain is controlled by an AGC voltage.

In this experiment, you will build various JFET circuits to see how the JFET can act as a switch, voltage-variable resistance, and AGC device.

Required Reading

Chapter 13 (Secs. 13-8 and 13-9) of *Electronic Principles*, 6th ed.

Equipment

1 audio generator
2 power supplies: adjustable to ±15 V
1 diode: 1N4148 or 1N914
3 JFETs: MPF102 (or equivalent)
4 $\frac{1}{2}$-W resistors: 2.2 kΩ, 10 kΩ, two 100 kΩ
2 capacitors: 1 μF
1 switch: SPST
1 VOM (analog or digital multimeter)
1 oscilloscope

Procedure

ANALOG SWITCH

1. Measure the approximate R_{DS} of each JFET as follows. Short the gate and source together. Connect the positive lead of an ohmmeter to the drain, and the negative lead to the source. Record the values of R_{DS} in Table 34-1. Throughout this experiment, you may use R_{DS} as an approximation for r_{ds}.
2. In Fig. 34-1, calculate v_{out} for each JFET when V_{GS} is zero. Record your answers in Table 34-1.
3. Connect the circuit with the ac signal source at 100 mV pp and 1 kHz.
4. Measure and record the ac output voltage with S_1 open and S_1 closed.
5. Repeat Step 4 for the other two JFETs.

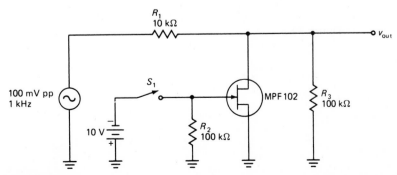

Figure 34-1

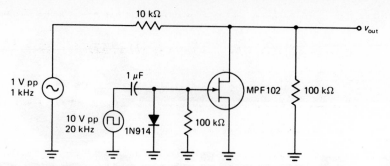

Figure 34-2

JFET CHOPPER

6. What do you think the output voltage of Fig. 34-2 will look like? Sketch the expected waveform in Table 34-2.

7. Connect the circuit with the specified ac input voltages and frequencies.

8. Look at the output voltage. Set the trigger select of the oscilloscope to the 1-kHz signal. Slowly vary the 1-kHz frequency until you see a steady chopped waveform. Sketch the waveform in Table 34-2.

VOLTAGE-VARIABLE RESISTANCE

9. As V_{GG} is varied from zero to a value more negative than $V_{GS(\text{off})}$, the peak-to-peak output voltage of Fig. 34-3 will change. Do you think that it will increase or decrease?

10. Connect the circuit.

11. Set up each value of V_{GG} shown in Table 34-3. Measure and record the ac output voltage. Calculate and record the value of r_{ds}.

AGC CIRCUIT

12. Connect the circuit of Fig. 34-4. Adjust V_{GG} to get maximum output signal. Measure and record v_{out} and V_{GG} in Table 34-4.

13. Adjust V_{GG} until v_{out} drops in half. Measure and record v_{out} and V_{GG}.

14. Repeat Step 13 two times.

TROUBLESHOOTING

15. Assume that V_{GG} is set to produce maximum output in Fig. 34-4. For each set of dc and ac symptoms listed in Table 34-5, figure out what the corresponding trouble may be. Insert your suspected trouble, then check the dc and ac voltages. When you locate each trouble, record it in Table 34-5.

CRITICAL THINKING

16. Select a value of R_1 in Fig. 34-1 that increases the attenuation by a factor of 10 when S_1 is open. Connect and test the circuit with your design value.

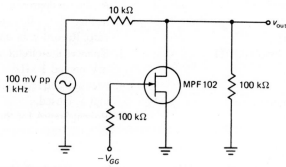

Figure 34-3

17. Repeat Steps 1 to 14 using EWB or an equivalent circuit simulator. Do not record any new values. But make sure that you get reasonable agreement be-

tween the EWB measurements and the values recorded earlier.

18. If you are using the CD-ROM version of this book, click on the Assignments menu and select Chap. 13.

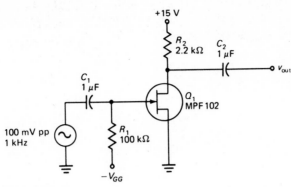

Figure 34-4

Data for Experiment 34

TABLE 34-1. JFET ANALOG SWITCH

JFET	Measured R_{DS}	Calculated v_{out}	Measured v_{out} S_1 open	S_1 closed
1				
2				
3				

TABLE 34-2. JFET CHOPPER

Expected waveform (sketch below)	Experimental waveform

TABLE 34-3. VOLTAGE-VARIABLE RESISTANCE

V_{GG}	v_{out}	r_{ds}
0		
−0.5 V		
−1 V		
−1.5 V		
−2 V		
−2.5 V		
−3 V		
−3.5 V		
−4 V		
−4.5 V		
−5 V		

TABLE 34-4. AGC CIRCUIT

Condition	v_{out}	V_{GG}
Maximum output		
Max/2		
Max/4		
Max/8		

TABLE 34-5. TROUBLESHOOTING

DC Symptoms		AC Symptoms		Trouble
V_G	V_D	v_g	v_d	
OK	OK	0	0	
OK	0	OK	0	
0	0	0	0	

Questions for Experiment 34

1. The JFET analog switch of Fig. 34-1 attenuates the signal when: (   )
   (a) it is on the negative half-cycles;   **(b)** S_1 is open;   **(c)** S_1 is closed;   **(d)** R_1 is shorted.
2. The gate circuit of Fig. 34-2 contains a: (   )
   (a) 1-kHz frequency;   **(b)** positive dc clamper;   **(c)** negative dc clamper;   **(d)** forward bias.
3. With the voltage-variable resistance of Fig. 34-3, the output signal increases when V_{GS}:   (   )
   (a) becomes more negative;   **(b)** stays the same;   **(c)** becomes more positive;   **(d)** is zero.
4. In the AGC circuit of Fig. 34-4, the output voltage decreases when: (   )
   (a) the ac input signal increases;   **(b)** V_{GG} goes to zero;   **(c)** V_{GG} becomes more negative;   **(d)** none of the foregoing.
5. Explain how the JFET analog switch of Fig. 34-1 works.

6. Explain how the voltage-variable resistance circuit of Fig. 34-3 works.

TROUBLESHOOTING

7. The dc drain voltage is zero in Fig. 34-4. Name three possible troubles.

8. The output voltage of Fig. 34-1 is zero with S_1 open or closed. Name three possible troubles.

CRITICAL THINKING

9. Which do you think is probably better for a JFET analog switch: a low or high $r_{ds(on)}$ when $V_{GS} = 0$? Explain your reasoning.

10. Optional. Instructor's question.

150

Power FETs

The MOSFET has an insulated gate that results in extremely high input impedance. The depletion-type MOSFET, also called a normally on MOSFET, can operate in either the depletion or enhancement mode. The enhancement-type MOSFET, also called a normally off MOSFET, can operate only in the enhancement mode. Power FETs are enhancement-type MOSFETs that are useful in applications requiring high load power, including audio amplifiers, RF amplfiers, and interfacing. In this experiment you will connect some circuits to get a better understanding of how they work.

Required Reading

Chapter 14 (Secs. 14-2 to 14-6) of *Electronic Principles*, 6th ed.

Equipment

1 audio generator
1 power supply: 15 V
1 diode: L53RD or similar red LED
1 power FET: Radio Shack 276-2072 (same as IFR510) or VN10KM
5 ½-W resistors: 330 Ω, 560 Ω, 1 kΩ, two 22 kΩ
2 capacitors: 0.1 μF, 100 μF (16-V rating or better)
1 switch: SPST
2 potentiometers: 1 kΩ, 5 kΩ (or similar low-resistance pots)
1 VOM (analog or digital multimeter)
1 oscilloscope
Graph paper

Procedure

THRESHOLD VOLTAGE

1. Connect the circuit of Fig. 35-1a. (See Fig. 35-1b for connections.)
2. Most data sheets define threshold voltage as the gate voltage that produces a drain current of 10 μA. Adjust the coarse and fine controls to get a drain current of 10 μA.
3. Record the threshold voltage in Table 35-1.

TRANSCONDUCTANCE CURVE

4. Set the fine control to maximum. Adjust the coarse control to produce a drain current slightly more than 10 mA. Adjust the fine control to get each drain current listed in Table 35-2. Record each gate voltage.
5. Graph your data, I_D versus V_{GS}.

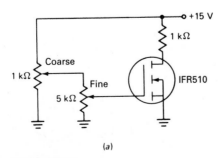
(a)

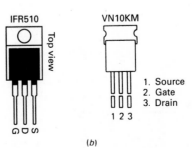
(b)

Figure 35-1

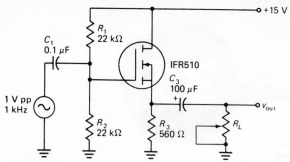

Figure 35-2

VOLTAGE-DIVIDER BIAS

6. Assume a V_{GS} of +2 V in Fig. 35-2. Calculate I_D, V_G, and V_S. Record your answers in Table 35-3.
7. Connect the circuit.
8. Measure and record I_D, V_G, and V_S.

SOURCE FOLLOWER

9. Calculate the g_m of the transistor in Fig. 35-2. (Use your graph for this.) Calculate the voltage gain and output impedance for a load resistance of infinity (R_L open). Record your answers in Table 35-4.
10. Connect the circuit with an infinite R_L. Measure and record the voltage gain for an input of 1 V pp. Also measure and record the MPP value.
11. Measure and record the output impedance using the matched-load method (described in Experiment 33).

DRIVING AN LED

12. Assume the R_{DS} of the power FET is 5 Ω. Calculate the I_D and V_D in Fig. 35-3 with S_1 closed. Record your answers in Table 35-5.
13. Connect the circuit. Open and close S_1. The LED should light when S_1 is closed and go out when S_1 is open.
14. Close S_1. Measure I_D and V_D.
15. Record in Table 35-5.

TROUBLESHOOTING

16. Table 35-6 lists some symptoms for the circuit of Fig. 35-3. Try to figure out the corresponding trouble. Insert the suspected trouble and verify the symptoms. Record each trouble.

CRITICAL THINKING

17. Select a value of current-limiting resistance in Fig. 35-3 that produces an LED current of approximately 20 mA.
18. Connect the circuit with your design value. Measure the LED current. Record your data here:

$R = $ _____

$I = $ _____

APPLICATION (OPTIONAL)—SCR-CONTROLLED MOTOR

19. Connect the circuit of Fig. 35-4 using a dc motor such as the Radio Shack RS273-256 (used earlier in Experiment 7). Do not apply power at this time.
20. Firmly attach a piece of electrical tape to the motor shaft, leaving a ½-inch flap. The tape will allow you to see the direction of rotation at slower speeds. Now, apply the +12 V of supply voltage. If the motor is running, vary the potentiometer until it stops.
21. Measure V_{GS} while varying the 1-MΩ potentiometer. Record the voltages at which the motor turns on and it turns off.
22. Slowly vary the potentiometer until the motor turns slowly enough for you to see the direction of rotation.
23. Reverse the leads on the motor. Again, slowly vary the potentiometer until the motor turns slowly enough for you to see the direction of rotation.
24. What did you learn in Steps 21 to 23?

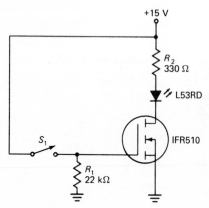

Figure 35-3

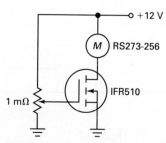

Figure 35-4

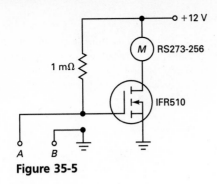

Figure 35-5

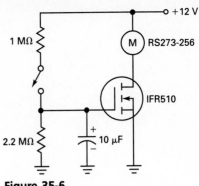

Figure 35-6

MOISTURE-SENSING CIRCUIT

25. Connect the circuit of Fig. 35-5.
26. The motor will run as soon as power is applied to the circuit.
27. Short points A and B. The motor will stop. Open the points, and the motor will run again.
28. If a small container of water is available, dip points A and B into the water. The motor will stop. Remove the points from the water, and the motor will run again. What did you learn in Steps 26 to 28?

SOFT TURN-ON CIRCUIT

29. Connect the circuit of Fig. 35-6. The motor should not be running when the switch is open.

30. Calculate and record the charging time constant when the switch is closed:

_____

31. Connect a voltmeter across the capacitor. Then, close the switch and watch the voltmeter as it measures the increasing voltage across the capacitor. At some point the motor will turn on.
32. Remove the power from the circuit. Replace the 10 μF by 100 μF. Then, repeat Steps 30 and 31.
33. What did you learn in Steps 30 to 32?

COMPUTER (OPTIONAL)

34. Repeat Steps 1 to 18 using EWB or an equivalent circuit simulator.
35. If you are using the CD-ROM version of this book, click on the Assignments menu and select Chap. 14.

Data for Experiment 35

TABLE 35-1. THRESHOLD VOLTAGE

$V_{GS(th)} =$

TABLE 35-2. TRANSCONDUCTANCE CURVE

I_D	V_{GS}
10 μA	
0.5 mA	
1 mA	
2 mA	
3 mA	
4 mA	
5 mA	
6 mA	
7 mA	
8 mA	
9 mA	
10 mA	

TABLE 35-3. VOLTAGE-DIVIDER BIAS

	Calculated			Measured	
I_D	V_G	V_S	I_D	V_G	V_S

TABLE 35-4. SOURCE FOLLOWER

Calculated $g_m =$	
Calculated $A =$	
Calculated $r_{out} =$	
Measured $A =$	
Measured PP $=$	
Measured $r_{out} =$	

TABLE 35-5. DRIVING A LOAD

	Calculated		Measured	
I_D	V_D	I_D		V_D

TABLE 35-6. TROUBLESHOOTING

S_1	V_D	LED	Trouble
Closed	0 V	out	
Closed	$\cong +14$ V	out	
Open	0 V	on	

Questions for Experiment 35

1. The threshold voltage of the power FET was the gate voltage that produced a drain current of: ( )
   (a) 10 μA;   (b) 100 μA;   (c) 1 mA;   (d) 10 mA.

2. When the drain current is approximately 10 mA, the transconductance of the power FET used in this experiment was closest to: ( )
   (a) 100 μS;   (b) 1000 μS;   (c) 2500 μS;   (d) 30,000 μS.

3. The calculated drain current in Table 35-3 is approximately: ( )
   (a) 1.1 mA;   (b) 5.26 mA;   (c) 9.82 mA;   (d) 12.5 mA.

4. The voltage gain in Table 35-4 is: ( )
   (a) less than 0.5;   (b) slightly less than unity;   (c) greater than unity;   (d) around 10.

5. The measured drain voltage of Table 35-5 was: ( )
   (a) large;   (b) less than 1 V;   (c) equal to supply voltage;   (d) unstable.

6. Describe what the power FET does in Fig. 35-3.

TROUBLESHOOTING

7. What is the last trouble you recorded in Table 35-6? Why does it produce the given symptoms?

8. Suppose the dc voltage across the 560-Ω resistor of Fig. 35-2 is zero. Name three possible troubles.

CRITICAL THINKING

9. What value of resistance did you record in Step 18? How did you arrive at this resistance?

The Silicon Controlled Rectifier

The silicon controlled rectifier (SCR) acts like a normally off switch. To turn it on, you have to apply a trigger to the gate. Once on, the SCR acts like a closed switch. You can then remove the trigger and the SCR remains closed. The only way to open the SCR is to reduce the supply voltage to a low value near zero.

Required Reading

Chapter 15 (Secs. 15-1 to 15-3) of *Electronic Principles*, 6th ed.

Equipment

1 power supply: adjustable from approximately 0 to 15 V with current limiting
1 red LED: L53RD or equivalent
2 transistors: 2N3904, 2N3906
1 SCR: Radio Shack 276-1067 or equivalent
6 ½-W resistors: two 330 Ω, 470 Ω, two 1 kΩ, 10 kΩ
1 potentiometer: 1 kΩ
1 VOM (analog or digital multimeter)
1 oscilloscope

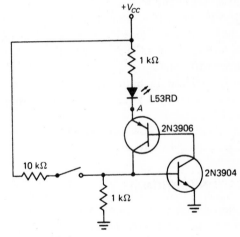

Figure 36-1

Procedure

TRANSISTOR LATCH

1. The transistor latch of Fig. 36-1 simulates an SCR. Assume that the LED of Fig. 36-1 is off. For a V_{CC} of +15 V, calculate the voltage between point A and ground. Record in Table 36-1. Also calculate and record the LED current.

2. Assume the switch of Fig. 36-1 is momentarily closed, then opened. Calculate and record the voltage at point A to ground for a V_{CC} of +15 V. Also calculate and record the LED current.

3. Connect the circuit with the switch open and a V_{CC} of +15 V.

4. The LED should be out. If not, reduce the supply voltage to zero, then back to +15 V.

5. With the LED off, measure and record the voltage at point A and the LED current.

6. Close the switch. The LED should come on.

7. Open the switch. The LED should stay on.

8. With the LED on, measure and record the voltage at point A. Measure and record the LED current. (When you break the circuit to insert the ammeter, the LED will go out. Close the switch to turn on the LED and measure the current.)

9. With the LED on, reduce the supply voltage until the LED goes off. Then increase the supply voltage and notice that the LED remains off.

10. Close the switch, then open it. The LED should be on.

SCR CIRCUIT

11. In Fig. 36-2, assume that V_{CC} is +15 V and that the LED is off. Calculate and record V_2 in Table 36-2. Also calculate and record the LED current.

12. Assume the LED is on. Calculate and record V_2 and I_{LED}.

157

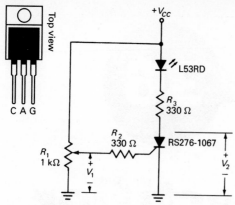

Figure 36-2

Figure 36-4

13. Assume a gate trigger current of 7 mA and a gate trigger voltage of 0.75 V. If the LED is off, what is the value of V_1 needed to turn on the LED? Record your answer in Table 36-2.

14. Assume a holding current of 6 mA. Calculate and record the value of V_{CC} where SCR turns off.

15. Connect the circuit with R_1 reduced to zero.

16. Adjust V_{CC} to +15 V. The LED should now be out. (If the LED is on, reduce V_{CC} to zero, then increase it back to +15 V. The LED should now be out.) Measure and record V_2 and I_{LED}.

17. Slowly increase V_1 until the LED just comes on. Measure and record V_2 and I_{LED}. Also measure and record V_1.

18. Reduce V_1 to zero. Slowly decrease V_{CC} until the LED just goes out. Measure and record V_{CC}.

APPLICATION (OPTIONAL)

19. Connect the crowbar circuit of Fig. 36-3.

20. Connect a dc voltmeter across the 1 kΩ. It should read approximately 5 V.

21. Slowly increase the supply voltage while watching the voltmeter reading. The crowbar should activate somewhere above 6.2 V. When it does, the load voltage will drop to a low value.

22. Reduce the supply voltage to zero. Then, repeat Step 21. What did you learn about a crowbar?

23. What did you learn in Steps 21 and 22?

24. Connect the motor-control circuit of Fig. 36-4 using a small dc motor such as the Radio Shack RS276-1067 (9 to 18 V). Do not apply power at this time.

25. Firmly attach a piece of electrical tape to the motor shaft, leaving a ½-inch flap.

26. Now, apply power and vary the 5-kΩ resistance in both directions over its entire range. The motor speed should change from maximum to minimum, and vice versa. (*Note:* If you cannot turn the motor off, add 4.7 kΩ in series with the variable resistance.)

27. Vary the resistance slowly and notice that you can control the speed of the motor from very slow to full speed.

28. Reverse the motor leads and repeat Steps 26 and 27.

29. What did you learn in Steps 26 to 28?

COMPUTER (OPTIONAL)

30. Repeat Steps 1 to 18 using EWB or an equivalent circuit simulator. Do not record any new values. But make sure that you get reasonable agreement between the EWB measurements and the values recorded earlier.

31. If you are using the CD-ROM version of this book, click on the Assignments menu and select Chap. 15.

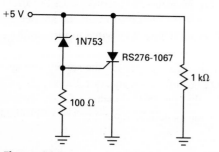

Figure 36-3

Data for Experiment 36

TABLE 36-1. TRANSISTOR LATCH

	Calculated		Measured	
LED	V_A	I_{LED}	V_A	I_{LED}
Off				
On				

TABLE 36-2. SCR CIRCUIT

LED off:
  Calculated V_2 = _____

  Calculated I_{LED} = _____

  Measured V_2 = _____

  Measured I_{LED} = _____

LED on:
  Calculated V_2 = _____

  Calculated I_{LED} = _____

  Measured V_2 = _____

  Measured I_{LED} = _____

Triggering:
  Calculated V_1 = _____

  Measured V_1 = _____

Holding:
  Calculated V_{CC} = _____

  Measured V_{CC} = _____

Questions for Experiment 36

1. When the LED of Fig. 36-1 is on, the voltage from point A to ground is closest to: ( )
   (a) 0;   (b) 3 V;   (c) 10 V;   (d) V_{CC}.
2. After the LED of Fig. 36-1 comes on, the current through it is approximately: ( )
   (a) 0;   (b) 5 mA;   (c) 9 mA;   (d) 13 mA.
3. When the switch of Fig. 36-1 is closed, the voltage across the lower 1 kΩ is: ( )
   (a) 0;   (b) 0.7 V;   (c) 1.5 V;   (d) 15 V.
4. When the switch of Fig. 36-1 is closed, the current through the 2N3904 is closest to: ( )
   (a) 0;   (b) 5 mA;   (c) 8 mA;   (d) 14 mA.
5. After the LED of Fig. 36-2 comes on, the only way to make it go off is to: ( )
   (a) reduce V_1 to zero;   (b) increase V_1 to 15 V;   (c) reduce V_{CC} toward zero;   (d) increase V_{CC} to 15 V.
6. The calculated V_1 of Table 36-2 needed for triggering is closest to: ( )
   (a) 1 V;   (b) 1.9 V;   (c) 3.06 V;   (d) 4.78 V.

7. Assume a LED voltage of 2 V in Fig. 36-2. The current through the SCR when it   (   )
is conducting is closest to:

    (a) 0;   **(b)** 20 mA;   **(c)** 40 mA;   **(d)** 75 mA.

8. When V_{CC} is 15 V and the LED is off in Fig. 36-2, V_2 is equal to: (   )

    (a) 1 V;   **(b)** 1.9 V;   **(c)** 3.06 V;   **(d)** 15 V.

9. What is meant by the term "latch" as it applies to the SCR?

10. After an SCR is turned on and anode current is flowing, how do you turn it off?

11. Optional. Instructor's question.

Frequency Effects

A n ac amplifier is normally operated in some middle range of frequencies where its voltage gain is approximately constant. Below this midrange, the response is down 3 dB at the lower cutoff frequency. Above the midrange, the response is down 3 dB at the upper cutoff frequency.

Decibels are useful for measuring and specifying the frequency response of amplifiers. When using Bode plots, we normally plot the voltage gain in decibels versus frequency on semilog paper. This allows us to see the frequency response over several decades.

Risetime is a useful way to specify the response of an amplifier. As discussed in the textbook, you can measure the risetime of an amplifier and calculate its upper cutoff frequency by using Eq. (16-29) in the textbook.

Required Reading

Chapter 16 of *Electronic Principles*, 6th ed.

Equipment

1 power supply: adjustable to 10 V
1 VOM (analog or digital multimeter)
1 transistor: 2N3904
5 ½-W resistors: 1 kΩ, 2.2 kΩ, 3.9 kΩ, 8.2 kΩ, 10 kΩ
7 capacitors: 220 pF, 470 pF, 0.1 μF, 0.47 μF, 10 μF, 470 μF
1 oscilloscope

Procedure

1. Answer the following questions for capacitors C_1 to C_4 in Fig. 37-1.
   Which are the capacitors that influence the lower cutoff frequency?

   Answer = _____ .
   Which is the capacitor that affects the upper cutoff frequency?

   Answer = _____ .

2. Connect the circuit of Fig. 37-1 using the following capacitances: $C_1 = 0.1$ μF, $C_2 = 10$ μF, $C_3 = 470$ μF, and $C_4 = 470$ pF. Capacitance C_4 is included to show the effects of stray wiring and other capacitances across the load.

3. Set the frequency of the input sine wave to 10 kHz. Use channel A of the oscilloscope to measure peak-to-peak voltage across the audio generator (adjust to 20 mV pp). Use channel B to measure the peak-to-peak voltage across the final load (8.2 kΩ). What is the voltage gain from the audio generator to the final output? What is this voltage gain in decibels?

   $A =$ _____ . $A_{db} =$ _____ .

4. Vary the frequency of the input as needed to locate the lower cutoff frequency. Record its value here:

   $f_1 =$ _____ .

5. Measure and record the voltage gain one decade below the cutoff frequency recorded in Step 4.

   $A_{db} =$ _____ .

6. In this circuit, C_1 is producing the dominant cutoff frequency. Change C_1 from 0.1 μF to 0.47 μF. Estimate the new value of the lower cutoff frequency.

   Answer = _____ .

7. Repeat Step 4:

   $f_1 =$ _____ .

8. Vary the frequency of the input as needed to locate the upper cutoff frequency. Record its value here:

   $f_2 =$ _____ .

9. Change C_4 from 470 pF to 220 pF. Estimate the new value of the upper cutoff frequency.

   Answer = _____ .

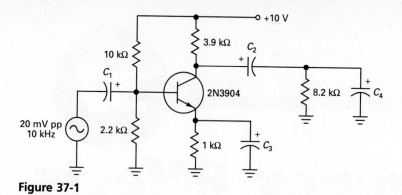

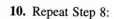

Figure 37-1

10. Repeat Step 8:

$f_2 =$ _____ .

11. Your wiring of the circuit may have some effect on the upper cutoff frequency. Can you explain why this is so? Use the space below:

12. If semilog paper is available, draw an ideal Bode plot of voltage gain versus frequency for the original circuit. Assume a 20-dB rolloff on both the low and the high side. If semilog paper is not available, sketch the ideal Bode plot and label the important parts of the graph.

13. Restore the original circuit of Fig. 37-1 by making $C_1 = 0.1 \ \mu F$ and $C_4 = 470 \ pF$. This time, use a square-wave input signal with a peak-to-peak value of 20 mV and a frequency of 10 kHz. Look at the output signal. Sketch the waveform here:

14. The normally horizontal lines of the square wave are now sloped at an angle. This effect is called *sag*. The charging and discharging of the dominant coupling capacitor C_1 are causing this sag. Change C_1 from 0.1 μF to 0.47 μF and notice how the sag decreases.

15. Increase the frequency of the input square wave to 50 kHz. Measure the risetime and record the value here:

$T_R =$ _____ .

16. Calculate the upper cutoff frequency and record its value here:

$f_2 =$ _____ .

This value should be in reasonable agreement with the value recorded in Step 8.

17. Change C_4 from 470 pF to 220 pF. Estimate the new risetime.

$T_R =$ _____ .

18. Repeat Step 15:

$T_R =$ _____ .

19. Remove C_4. Measure and record the risetime.

$T_R =$ _____ .

Questions for Experiment 37

1. If C_1 is dominant, an increase in C_1 has what effect on the lower cutoff frequency? ( )
   (a) decreases it; (b) doubles it; (c) increases it; (d) no effect.

2. What effect does a decrease in C_4 have on the upper cutoff frequency? ( )
   (a) decreases it; (b) doubles it; (c) increases it; (d) no effect.

3. If C_1 is dominant, an increase in C_1 has what effect on the sag? ( )
   (a) decreases it; (b) doubles it; (c) increases it; (d) no effect.

4. What effect does a decrease in C_4 have on the risetime? ( )
   (a) decreases it; (b) doubles it; (c) increases it; (d) no effect.

5. In this experiment, the decibel voltage gain was closest to: ( )
   (a) 0 dB; (b) 20 dB; (c) 40 dB; (d) 60 dB.

6. If you want to get minimum risetime, the stray-wiring capacitance across the load ( )
   must be:
   (a) minimum; (b) maximum; (c) equal to the output coupling
   capacitor; (d) none of these.

7. To measure risetime, the input signal must be a: ( )
   (a) sine wave; (b) square wave; (c) triangular wave; (d) none of
   these.

8. If the voltage gain of an amplifier is 10, the voltage gain at the upper cutoff ( )
   frequency is:
   (a) −3 dB; (b) 7 dB; (c) 13 dB; (d) 7.07.

9. Optional. Instructor's question.

10. Optional. Instructor's question.

The Differential Amplifier

T he differential amplifier is the direct-coupled input stage of the typical op amp. The most common form of a diff amp is the double-ended input and single-ended output circuit. Some of the important characteristics of a diff amp are the input offset current, input bias current, input offset voltage, and common-mode rejection ratio. In this experiment you will build a diff amp and measure the foregoing quantities.

Required Reading

Chapter 17 (Secs. 17-1 to 17-5) of *Electronic Principles*, 6th ed.

Equipment

1   audio generator
2   power supplies: ± 15 V
10 ½-W resistors: two 22 Ω, two 100 Ω, two 1.5 kΩ, two 4.7 kΩ, two 10 kΩ (5% tolerance)
2   transistors: 2N3904
1   capacitor: 0.47 μF
1   VOM (analog or digital multimeter)
1   oscilloscope

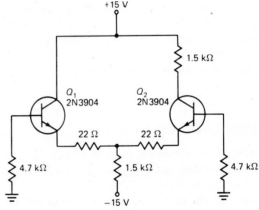

Figure 38-1

Procedure

TAIL CURRENT AND BASE CURRENTS

1. Notice the pair of swamping resistors (22 Ω) in Fig. 38-1. These have to be included in this experiment to improve the match between the discrete transistors. In Fig. 38-1, you may assume the typical h_{FE} is 200. Calculate the approximate tail current. Record in Table 38-1. Also calculate and record the base current in each transistor.
2. Connect the circuit of Fig. 38-1.
3. Measure and record the tail current.
4. Use the VOM as an ammeter to measure the base current in each transistor. If your VOM is not sensitive enough to measure microampere currents, then use the oscilloscope on dc input to measure the voltage across

each base resistor and calculate the base current. Record the base currents in Table 38-1.

INPUT OFFSET AND BIAS CURRENTS

5. With the calculated data of Table 38-1, calculate the values of input offset current and input bias current. Record your theoretical answers in Table 38-2.
6. With the measured data of Table 38-1, calculate the values of $I_{\text{in(off)}}$ and $I_{\text{in(bias)}}$. Record your experimental answers in Table 38-2.

OUTPUT OFFSET VOLTAGE

7. In Fig. 38-2, assume that the base of Q_1 is grounded by a jumper wire. If both transistors are identical and

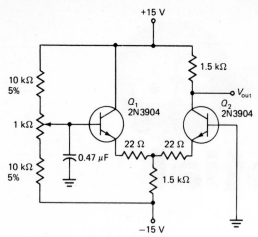

Figure 38-2

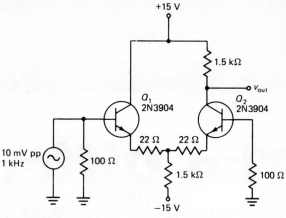

Figure 38-3

all components have the values shown, then the dc output voltage would have a value of approximately +7.85 V. For this part of the experiment, any deviation from +7.85 V is called *output offset voltage,* designated $V_{out(off)}$.

8. Connect the circuit of Fig. 38-2. Ground the base of Q_1 with a jumper wire. Measure the dc output voltage. Calculate the output offset voltage and record $V_{out(off)}$ in Table 38-3.

9. Remove the ground from the Q_1 base. Adjust the potentiometer until the output voltage is +7.85 V.

10. Measure the base voltage of Q_1. Record in Table 38-3 as $V_{in(off)}$.

DIFFERENTIAL VOLTAGE GAIN

11. Because of the swamping resistors in Fig. 38-3, the differential voltage gain is given by $R_C/2(r_E + r'_e)$. Calculate and record A in Table 38-4.

12. Connect the circuit. Set the audio generator at 1 kHz with a signal level of 10 mV pp.

13. Measure the output voltage. Calculate and record the experimental value of A.

COMMON-MODE VOLTAGE GAIN

14. Calculate the common-mode voltage gain of Fig. 38-3. Record A_{CM} in Table 38-4.

15. Put a jumper wire between the bases of your built-up circuit.

16. Increase the signal level until the output voltage is 0.5 V pp.

17. Measure the peak-to-peak input voltage. Calculate and record the experimental value of A_{CM}.

COMMON-MODE REJECTION RATIO

18. Calculate and record the theoretical value of CMRR using the calculated data of Table 38-4.

19. Calculate and record the experimental value of CMRR using the experimental data of Table 38-4.

TROUBLESHOOTING

20. In this part of the experiment, a collector-emitter short means that all three transistor terminals are shorted together. A collector-emitter open means the transistor is removed from the circuit.

21. In Fig. 38-3, estimate the dc output voltage for each trouble listed in Table 38-5.

22. Insert each trouble; measure and record the dc voltages of Table 38-5.

CRITICAL THINKING

23. Select resistance values in Fig. 38-3 to get a tail current of 3 mA and a dc output voltage of +7.5 V. Record the nearest standard values in Table 38-6.

24. Connect the circuit with your design values. Measure and record the tail current and dc output voltage.

COMPUTER (OPTIONAL)

25. Repeat Steps 1 to 24 using EWB or an equivalent circuit simulator. Do not record any new values. But make sure that you get reasonable agreement between the EWB measurements and the values recorded earlier.

26. If you are using the CD-ROM version of this book, click on the Assignments menu and select Chap. 17.

166

Data for Experiment 38

TABLE 38-1. TAIL AND BASE CURRENTS

Calculated	Measured
$I_T =$ _____	$I_T =$ _____
$I_{B1} =$ _____	$I_{B1} =$ _____
$I_{B2} =$ _____	$I_{B2} =$ _____

TABLE 38-2. INPUT OFFSET AND BIAS CURRENTS

Theoretical	Experimental
$I_{in(off)} =$ _____	$I_{in(off)} =$ _____
$I_{in(bias)} =$ _____	$I_{in(bias)} =$ _____

TABLE 38-3. INPUT AND OUTPUT OFFSET VOLTAGES

$V_{out(off)} =$ _____

$V_{in(off)} =$ _____

TABLE 38-4. VOLTAGE GAINS AND CMRR

Calculated	Experimental
$A =$ _____	$A =$ _____
$A_{CM} =$ _____	$A_{CM} =$ _____
CMRR $=$ _____	CMRR $=$ _____

TABLE 38-5. TROUBLESHOOTING

Trouble	Estimated V_{out}	Measured V_{out}
Q_1 CE short		
Q_1 CE open		
Q_2 CE short		
Q_2 CE open		

TABLE 38-6. CRITICAL THINKING

$R_E =$ _____

$R_C =$ _____

$I_T =$ _____

$V_{C2} =$ _____

Questions for Experiment 38

1. The tail current of Table 38-1 is closest to: ( )
   (a) 1 μA; (b) 23.8 μA; (c) 47.6 μA; (d) 9.53 mA.
2. The calculated base current of Fig. 38-1 is approximately: ( )
   (a) 1 μA; (b) 23.8 μA; (c) 47.6 μA; (d) 9.53 mA.
3. The calculated input bias current of Fig. 38-1 is approximately: ( )
   (a) 1 μA; (b) 23.8 μA; (c) 47.6 μA; (d) 9.53 mA.
4. The input offset voltage is the input voltage that removes the: ( )
   (a) tail current; (b) dc output voltage; (c) output offset voltage;
   (d) supply voltage.
5. The CMRR of Table 38-4 is closest to: ( )
   (a) 0.5; (b) 27.5; (c) 55; (d) 123.
6. Why is a high CMRR an advantage with a diff amp?

TROUBLESHOOTING

7. In Fig. 38-3, somebody mistakenly uses 150 Ω instead of 1.5 kΩ for the tail resistor. What are some of the dc and ac symptoms you can expect?

8. You are troubleshooting the circuit of Fig. 38-3. What ac voltage should an oscilloscope display at the junction of the 22-Ω resistor with respect to ground?

CRITICAL THINKING

9. What value did you use for R_E and R_C in your design? What is the new value of CMRR?

10. Optional. Instructor's question.

168

Experiment 39

Differential-Amplifier Supplement

The preceding experiment on the differential amplifier focused on a single-ended output. In this experiment, we will use a double-ended output. Recall the basic theory given in the textbook. A double-ended output has twice as much voltage gain as a single-ended output. Also, the outputs from the collectors have equal magnitudes but opposite phases.

Required Reading

Chapter 17 (Secs. 17-1 and 17-5) of *Electronic Principles,* 6th ed.

Equipment

1 audio generator
2 power supplies: ±10 V
1 VOM (analog or digital multimeter)
2 transistors: 2N3904
6 ½-W resistors: two 1 kΩ, one 4.7 kΩ, three 10 kΩ
1 potentiometer: 10 kΩ
1 oscilloscope

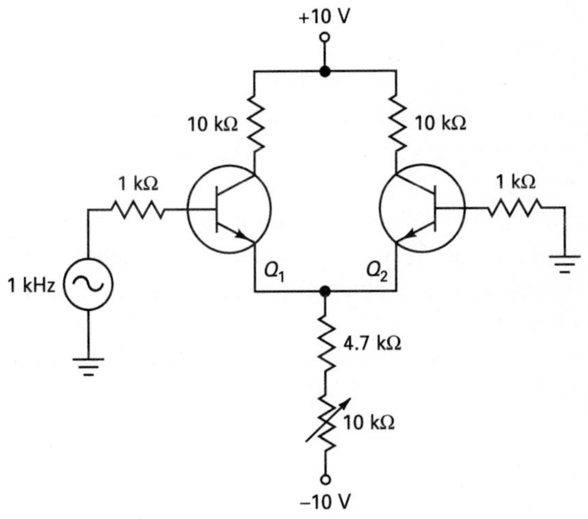

Figure 39-1

Procedure

SINGLE-ENDED INPUT AND DOUBLE-ENDED OUTPUT

1. Connect the circuit of Fig. 39-1.
2. Reduce the generator amplitude to zero. Then, measure the dc collector voltage of Q_2. Vary the emitter resistance until you have approximately 5 V on the Q_2 collector.
3. Measure and record the dc voltages listed in Table 39-1.
4. Use channel 1 of the oscilloscope to look at the ac base voltage of Q_1. Adjust the audio generator to get a frequency of 1 kHz and a sinusoidal amplitude of 50 mV peak-to-peak.
5. Use channel 2 to look at the ac collector voltage of Q_1. You should see an amplified and inverted sine wave. Record the peak-to-peak value and phase in Table 39-2. (*Note:* The phase is recorded as an example in this first measurement.)
6. Repeat Step 5 for Q_2 and the top of the tail resistor (where the two emitters connect).
7. Turn off the power and reconnect the circuit with the audio generator on the other side of the diff amp.
8. Repeat Steps 5 and 6. (*Note:* Use Table 39-3.)

DIFFERENTIAL MODE

9. Connect the circuit of Fig. 39-2. The first stage is called a *phase splitter,* a circuit that ideally produces equal-magnitude and opposite-phase signals.

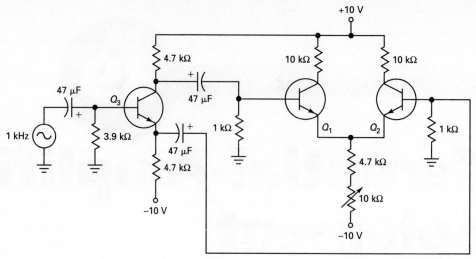

Figure 39-2

10. Turn on the power. Repeat Step 2 to check that the dc collector voltage on Q_2 is still approximately 5 V.

11. Adjust the audio generator to get 1 kHz and a sinusoidal amplitude of 50 mV pp at the base of Q_1. Table 39-4 shows the phase of v_1 as 0° because this signal will be used as a reference for other measurements.

12. Look at the ac collector voltage of Q_1. You should see an amplified and inverted sine wave. Record the peak-to-peak value and phase in Table 39-4.

13. Repeat Step 12 for the other voltages listed in Table 39-4. *Note:* When measuring v_{out}, use the oscilloscope in the difference mode with channel 1 measuring v_{C2} and channel 2 measuring v_{C1}. If your oscilloscope does not include the difference function, mentally calculate the difference between v_{C2} and v_{C1}.

14. Use the data of Table 39-4 to calculate the voltage gain of Q_1 and Q_2:

Gain of Q_1 = _____ .

Gain of Q_2 = _____ .

COMMON-MODE OPERATION

15. Turn off the power and connect the circuit of Fig. 39-3.

16. Apply power. Measure and record all quantities listed in Table 39-5.

17. Use the data of Table 39-5 to calculate the voltage gain of Q_1 and Q_2:

Gain of Q_1 = _____ .

Gain of Q_2 = _____ .

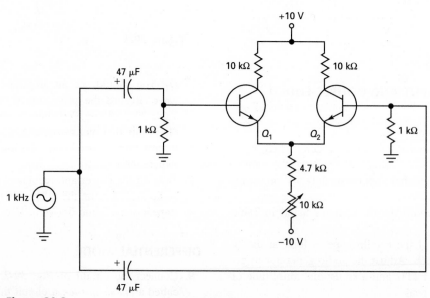

Figure 39-3

170

Data for Experiment 39

TABLE 39-1. DC VOLTAGES

V_{C1}	V_{C2}	V_E	V_{EE}	V_{B1}	V_{B2}

TABLE 39-2. NONINVERTING INPUT

	v_1	v_{c1}	v_{c2}	v_e
Magnitude	50 mV			
Phase	0°	180°		

TABLE 39-3. INVERTING INPUT

	v_2	v_{c1}	v_{c2}	v_e
Magnitude	50 mV			
Phase	0°			

TABLE 39-4. DIFFERENTIAL INPUT

	v_1	v_{c1}	v_2	v_{c2}	v_e	v_{out}
Magnitude	50 mV					
Phase	0°					

TABLE 39-5. COMMON-MODE INPUT

	v_1	v_{c1}	v_2	v_{c2}	v_e	v_{out}
Magnitude	50 mV					
Phase	0°					

Questions for Experiment 39

1. In Table 39-2, compare and explain the phase of v_{c1} and v_{c2} with the reference phase of v_1:

2. In Table 39-3, compare and explain the phase of v_{c1} and v_{c2} with the reference phase of v_2:

3. With the data of Table 39-2, calculate the single-ended voltage gain of each transistor. Also, calculate the differential voltage gain. Briefly show your calculations:

4. With the data of Table 39-4, explain the significance of the waveforms at v_{c1} and v_{c2}.

5. With the data of Table 39-4, explain the significance of the waveform at v_e.

6. Compare the waveforms of Table 39-4 and Table 39-5. Now, explain why differential voltage gain is much greater than common-mode voltage gain.

Introduction to Op-Amp Circuits

When connecting op-amp circuits, you need to take the following precautions to avoid possible damage or other unwanted effects:

1. Adjust the supply voltages to the desired level. Then, turn the power off and connect the power supplies to the op-amp pins.
2. To avoid damaging an op amp, the power should remain off while you are bread-boarding the rest of the circuit.
3. Make sure that the ac source is turned off before connecting it to the op-amp circuit.
4. To avoid oscillations and other unwanted effects, make sure that all wiring is as short as possible.
5. After a final wiring check, you can turn on the dc supply voltages.
6. Next, you can turn on the ac source. To avoid possible damage to the op amp, the ac peak voltages should always be less than the supply voltages.
7. If oscillations appear, use decoupling capacitors on the supply pins. Manufacturers recommend capacitances between 0.1 and 1 μF between each supply pin and ground.
8. When powering down, reduce the ac source to zero. Be sure to do this before turning off the dc supplies.
9. The last step is to turn off the dc supply voltages.

Required Reading

Chapter 18 (Secs. 18-1 and 18-2) of *Electronic Principles,* 6th ed.

Equipment

1   function generator
2   power supplies: ± 15 V
1   op amp: 741C
5   ½-W resistors: two 100 Ω, 10 kΩ, two 100 kΩ
2   capacitors: 0.47 μF
2   potentiometers: 10 kΩ
1   VOM (analog or digital)
1   oscilloscope

Procedure

1. Connect the circuit of Fig. 40-1.
2. Use channel 1 of the oscilloscope to measure V_1, the dc voltage between the noninverting input (pin 3) and ground. *Note:* Use a sensitive range because the dc voltage will be less than ± 15 mV.
3. Use channel 2 of the oscilloscope to measure the dc output voltage (pin 6).
4. Vary the potentiometer. Somewhere near the middle of the range, you will see the dc output voltage change from positive to negative, or vice versa.
5. Slowly vary the potentiometer until the output goes into positive saturation. Record V_1 and V_{out} in Table 40-1.
6. Slowly vary the potentiometer until the output goes into negative saturation. Record V_1 and V_{out} in Table 40-1.

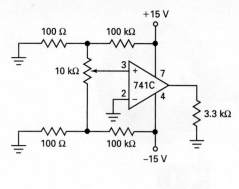

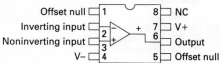

Figure 40-1

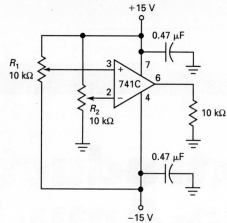

Figure 40-3

7. Repeat Steps 1 to 6 for Fig. 40-2. Record your data in Table 40-2.

8. The two circuits (Fig. 40-1 and Fig. 40-2) behave differently. What did you notice, and how can you account for the difference?

9. In Fig. 40-3, potentiometer R_2 can be adjusted to produce the V_2 voltages shown in Table 40-3. For each V_2 voltage, what is the V_1 voltage that causes the output to switch states? Record your answers under "Calculated V_1."

10. Build the circuit of Fig. 40-3. Adjust R_2 to get each V_2 voltage listed in Table 40-3. Vary R_1 until the output switches states. Record the V_1 values in Table 40-3.

11. Repeat Steps 9 and 10 for Fig. 40-4. Record your answers in Table 40-4.

12. Connect the circuit of Fig. 40-5 using the triangular output of a function generator.

13. Use channel 1 to look at the triangular input voltage.

14. Use channel 2 to look at the output voltage.

15. Adjust the amplitude of the triangular wave to 8 V peak to peak, and the frequency to 100 Hz.

16. Vary the potentiometer over a wide range and notice how the output duty cycle changes.

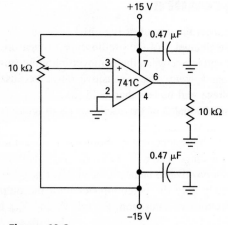

Figure 40-2

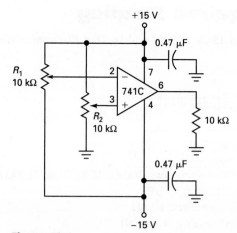

Figure 40-4

174

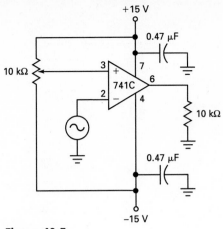

+15 V

0.47 μF

10 kΩ

3 +

7

741C

6

2 −

4

10 kΩ

0.47 μF

−15 V

Figure 40-5

17. Explain the results of Step 16. Why did the duty cycle change?

18. Reverse the two input signals by connecting the triangular input to pin 3 and the potentiometer to pin 2.
19. Repeat Steps 15 and 16.
20. Discuss the phase relations of Step 19.

Data for Experiment 40

TABLE 40-1. FINE-ADJUSTMENT CIRCUIT

	Positive Saturation	Negative Saturation
V_1		
V_{out}		

TABLE 40-2. COARSE-ADJUSTMENT CIRCUIT

	Positive Saturation	Negative Saturation
V_1		
V_{out}		

TABLE 40-3. NONINVERTING COMPARATOR

V_2	Calculated V_1	Measured V_1
0		
1		
2		
5		
10		

TABLE 40-4. INVERTING COMPARATOR

V_2	Calculated V_1	Measured V_1
0		
1		
2		
5		
10		

Questions for Experiment 40

1. In Fig. 40-1, the maximum positive voltage to pin 3 is: ( )
   (a) 0;   (b) 15 mV;   (c) 30 mV;   (d) 15 V.
2. In Fig. 40-2, the wiper of the potentiometer is much closer to the upper end. The ( )
   output of the op amp is closest to:
   (a) 0;   (b) 15 mV;   (c) 30 mV;   (d) 15 V.
3. In Fig. 40-3, R_2 is adjusted to get $V_2 = +7.5$ V. To switch the output state, the ( )
   resistance between the wiper and the upper end of the potentiometer must be
   approximately:
   (a) 0;   (b) 1 kΩ;   (c) 2.5 kΩ;   (d) 5 kΩ.   ( )

4. Decoupling capacitors are used on the supply pins of the op amp to prevent: ( )
  (a) output switching;   **(b)** oscillations;   **(c)** damaging the op amp;
  (d) saturation.

5. The data of Table 40-1 demonstrates that the voltage gain of an op amp is very: ( )
  (a) low;   **(b)** high;   **(c)** sharp;   **(d)** unreliable.

6. The data of Table 40-3 demonstrate that the output switches when the noninverting input voltage and the inverting input voltage are approximately: ( )
  (a) equal;   **(b)** 180° out of phase;   **(c)** dc voltages;   **(d)** ac voltages.

7. Explain why the duty cycle changes in a circuit like Fig. 40-5.

8. What do you consider to be the three most important things you learned in this experiment?

9. Optional. Instructor's question.

10. Optional. Instructor's question.

Inverting and Noninverting Amplifiers

Inverting and noninverting amplifiers are the most fundamental op-amp circuits. The closed-loop voltage gain of an inverting amplifier equals the ratio of the feedback resistance to the input resistance, R_2/R_1. On the other hand, the closed-loop voltage gain of a noninverting amplifier equals the ratio of the feedback resistance to the input resistance plus 1, $R_2/R_1 + 1$.

In this experiment, you will build both types of amplifiers and test their operation, and you will measure bandwidth and observe slew-rate distortion. You will also see how the MPP of an op amp decreases when the load resistance decreases.

Required Reading

Chapter 18 (Secs. 18-1 to 18-4) of *Electronic Principles,* 6th ed.

Equipment

1 audio generator
2 power supplies: ± 15 V
1 op amp: 741C
8 $\frac{1}{2}$-W resistors: two 100 Ω, three 1 kΩ, two 10 kΩ
1 potentiometer: 1 kΩ
1 VOM (analog or digital)
1 oscilloscope

Procedure

1. In Fig. 41-1, v_{in} is the voltage between the potentiometer wiper and ground. What is the output voltage in this circuit for each of the input voltages shown in Table 41-1? Record your answers under v_{out} (calculated).
2. Connect the circuit of Fig. 41-1.
3. To keep the wiring simple in this experiment, we are omitting the decoupling capacitors that are normally

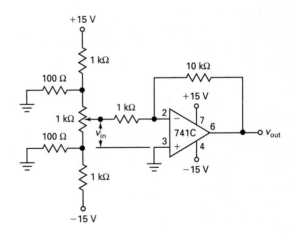

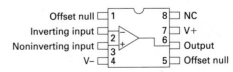

Figure 41-1

connected between the supply pins and ground. If the output signal is very noisy or other unwanted effects appear, add 0.47-μF capacitors to pins 4 and 7 as shown in Experiment 40.

4. Use channel 1 of the oscilloscope to measure v_{in}, and channel 2 to measure v_{out}. Vary the potentiometer over its full range while observing the input and output waveforms.

5. Adjust the potentiometer to produce each positive input voltage listed in Table 41-1. Measure and record the output voltage for each input voltage.

6. Table 41-2 shows negative input voltages for the circuit of Fig. 41-1. What are the theoretical output voltages? Record your answers in Table 41-2.

7. Measure and record the output voltage for each input voltage in Table 41-2.

8. Connect the circuit of Fig. 41-2. Use channels 1 and 2 of the oscilloscope to look at the input and output voltages.

9. Adjust the input voltage to 2 V pp and 1 kHz. Notice how the input and output waveforms are related in magnitude and phase.

10. Repeat Step 9 for the following frequencies: 10 Hz, 100 Hz, and 10 kHz. Summarize your observations about voltage gain and phase shift:

11. Repeat Step 9 for 100 kHz. Why does this happen?

12. Set the frequency to 1 kHz.

13. In Table 41-3, what is the peak-to-peak output voltage for each input voltage shown? Record your answers.

14. Measure the peak-to-peak output voltage for each input voltage of Table 41-3. Record your measured values.

15. Connect the circuit of Fig. 41-3. Use channels 1 and 2 of the oscilloscope to look at the input and output voltages.

16. Adjust the input voltage to 2 V pp and 1 kHz. Notice how the input and output waveforms are related in magnitude and phase.

17. In Table 41-4, what is the peak-to-peak output voltage for each input voltage shown? Record your answers.

18. Measure the peak-to-peak output voltage for each input voltage of Table 41-4. Record your measured values.

19. Adjust the generator amplitude to get an output voltage with a peak-to-peak value of 1 V. Increase the generator frequency until the output amplitude decreases to 0.7 V pp. Record the frequency as the bandwidth in Table 41-5. Is there any slew-rate distortion? Record *yes* or *no*.

20. Set the frequency to 1 kHz and adjust the generator amplitude to get an output voltage of 20 V pp. Increase the generator frequency until the output amplitude is 14 V pp. Record the frequency as the bandwidth. Is there any slew-rate distortion? Record in Table 41-5.

21. Reduce the generator amplitude to zero and adjust the frequency to 1 kHz.

22. Connect a load resistor of 10 kΩ between pin 6 and ground.

23. Increase the generator amplitude until clipping starts on the output signal. Measure and record the peak-to-peak output voltage as the MPP in Table 41-6.

24. Change the load resistance to 100 Ω. Vary the generator amplitude until clipping starts on the output signal. Record the MPP for this load resistance.

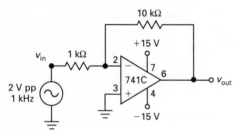

Figure 41-2

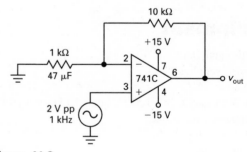

Figure 41-3

Data for Experiment 41

TABLE 41-1. INVERTING AMPLIFIER (POSITIVE INPUTS)

v_{in}	Calculated v_{out}	Measured v_{out}
0.1 V		
0.2 V		
0.5 V		
1 V		

TABLE 41-2. INVERTING AMPLIFIER (NEGATIVE INPUTS)

v_{in}	Calculated v_{out}	Measured v_{out}
−0.1 V		
−0.2 V		
−0.5 V		
−1 V		

TABLE 41-3. INVERTING AMPLIFIER (SINUSOIDAL INPUTS)

v_{in}	Calculated v_{out}	Measured v_{out}
0.1 V pp		
0.2 V pp		
0.5 V pp		
1 V pp		
2 V pp		

TABLE 41-4. NONINVERTING AMPLIFIER (SINUSOIDAL INPUTS)

v_{in}	Calculated v_{out}	Measured v_{out}
0.1 V pp		
0.2 V pp		
0.5 V pp		
1 V pp		
2 V pp		

TABLE 41-5. BANDWIDTH OF NONINVERTING AMPLIFIER

v_{out}	BW	Slew-Rate Distortion
1 V pp		
20 V pp		

TABLE 41-6. MAXIMUM PEAK-TO-PEAK OUTPUT

R_L	MPP
10 kΩ	
100 Ω	

Questions for Experiment 41

1. The input voltage dividers of Fig. 41-1 reduce the total voltage across the poten- ( )
   tiometer to approximately:
   (a) 15 mV;   (b) 1.5 V;   (c) 3 V;   (d) 15 V.
2. In Fig. 41-1, the closed-loop voltage gain is: ( )
   (a) 1;   (b) 10;   (c) 11;   (d) 100.
3. In Fig. 41-2, the peak-to-peak output voltage at low frequencies is: ( )
   (a) 1 V;   (b) 2 V;   (c) 10 V;   (d) 20 V.
4. In Fig. 41-2, the voltage between pin 2 and ground is approximately: ( )
   (a) 0 V;   (b) 1 V;   (c) 2 V;   (d) 20 V.
5. The closed-loop voltage gain of Fig. 41-3 is approximately: ( )
   (a) 1;   (b) 10;   (c) 11;   (d) 100.
6. If the 10-kΩ resistor opens in Fig. 41-3, the output voltage is: ( )
   (a) 0;   (b) 10 V;   (c) sinusoidal;   (d) rectangular.
7. In Fig. 41-3, the product of the closed-loop voltage gain and the small-signal band- ( )
   width is closest to:
   (a) 1 kHz;   (b) 20 kHz;   (c) 90 kHz;   (d) 1 MHz.
8. Why did slew-rate distortion appear in Step 11 and not in Step 10?

9. Why was the MPP smaller in Step 24 than in Step 23?

10. Optional. Instructor's question.

The Operational Amplifier

An operational amplifier, or op amp, is a high-gain dc amplifier usable from 0 to over 1 MHz (typical). By connecting external resistors to an op amp, you can adjust the voltage gain and bandwidth to your requirements. Whether troubleshooting or designing, you have to be familiar with the characteristics of an op amp. These include the input offset current, input bias current, input offset voltage, common-mode rejection ratio, MPP value, short-circuit output current, slew rate, and power bandwidth. In this experiment you will connect and test a basic op-amp circuit.

Required Reading

Chapter 18 (Secs. 18-1 and 18-2) of *Electronic Principles*, 6th ed.

Equipment

1 audio generator
2 power supplies: ±15 V
8 ½-W resistors: two 100 Ω, 1 kΩ, two 10 kΩ, 100 kΩ, two 220 kΩ
3 op amps: 741C
2 capacitors: 0.47 μF
1 VOM (analog or digital multimeter)
1 oscilloscope

Figure 42-1

Procedure

INPUT OFFSET AND BIAS CURRENTS

1. The 741C has a typical $I_{in(bias)}$ of 80 nA. Assume that this is the base current in each 220-kΩ resistor of Fig. 42-1. Calculate dc voltages at the noninverting and inverting inputs. Record in Table 42-1.
2. Connect the circuit of Fig. 42-1.
3. Measure the dc voltage at the noninverting input. Record in Table 42-1.
4. Measure and record the inverting input voltage.
5. Repeat Steps 1 to 4 for the other 741Cs.

6. With the measured data of Table 42-1, calculate the base currents, then the values of $I_{in(off)}$ and $I_{in(bias)}$. Record your answers in Table 42-2.

OUTPUT OFFSET VOLTAGE

7. Connect the circuit of Fig. 42-2. Note: Bypass capacitors are used on each supply voltage to prevent oscillations, discussed in Chap. 22 of your textbook. These capacitors should be connected as close to the IC as possible.
8. Measure the dc output voltage. Record this value as $V_{out(off)}$ in Table 42-3.
9. Repeat Step 8 for the other 741Cs.

183

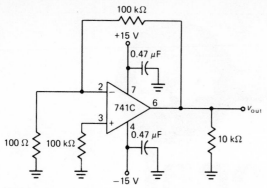

Figure 42-2

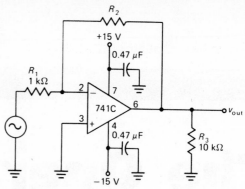

Figure 42-3

10. With the resistors shown in Fig. 42-2, the circuit has a voltage gain of 1000. Calculate the input offset voltage with

$$V_{\text{in(off)}} = \frac{V_{\text{out(off)}}}{1000}$$

Record your results in Table 42-3.

MAXIMUM OUTPUT CURRENT

11. Disconnect the right end of the 100-kΩ resistor from the output.

12. Connect the right end of the 100-kΩ resistor to the +15 V. This will apply approximately 15 mV to the inverting input, more than enough to saturate the op amp.

13. Replace the 10-kΩ load resistor by a VOM used as an ammeter. Since the ammeter has a very low resistance, it indicates the short-circuit output current.

14. Read and record I_{max} in Table 42-3.

15. Repeat Step 14 for the other 741Cs.

SLEW RATE

16. Connect the circuit of Fig. 42-3 with an R_2 of 100 kΩ.

17. Use the oscilloscope (time base around 20 μs/DIV) to look at the output of the op amp. Set the audio generator at 5 kHz. Adjust the signal level to get hard clipping on both peaks of the output signal (overdrive condition).

18. Measure the voltage change and the time change of the waveform. Calculate and record the slew rate in Table 42-4.

19. Repeat Step 18 for the other 741Cs.

POWER BANDWIDTH

20. Change R_2 to 10 kΩ. Set the ac generator at 1 kHz. Adjust the signal level to get 20 V pp out of the op amp.

21. Increase the frequency from 1 to 20 kHz and watch the waveform. Somewhere above 8 kHz, slew-rate distortion will become evident because the waveform will appear triangular and the amplitude will decrease.

22. Record the approximate (ballpark) frequency where slew-rate distortion begins (Table 42-4).

23. Repeat Steps 20 to 22 for the other 741Cs.

MPP VALUE

24. Set the ac generator at 1 kHz. Increase the signal level until clipping just starts on either peak.

25. Record the MPP in Table 42-4.

TROUBLESHOOTING

26. Measure the dc and ac output voltage for each trouble listed in Table 42-5.

27. Record your data in Table 42-5.

CRITICAL THINKING

28. As derived in your textbook, the voltage gain of a circuit like Fig. 42-3 is equal to R_2/R_1. Select a value of R_2 to get a voltage gain of 68.

29. Replace R_2 by your design value. Measure the voltage gain.

30. Record your design value and the measured voltage gain in Table 42-6.

COMPUTER (OPTIONAL)

31. Repeat Steps 1 to 30 using EWB or an equivalent circuit simulator. Do not record any new values. But make sure that you get reasonable agreement between the EWB measurements and the values recorded earlier.

32. If you are using the CD-ROM version of this book, click on the Assignments menu and select Chap. 18.

184

Data for Experiment 42

TABLE 42-1. INPUT VOLTAGES

Op Amp	Calculated		Measured	
	v_1	v_2	v_1	v_2
1				
2				
3				

TABLE 42-2. INPUT OFFSET AND BIAS CURRENTS

Op Amp	$I_{in(off)}$	$I_{in(bias)}$
1		
2		
3		

TABLE 42-3. INPUT AND OUTPUT OFFSET VOLTAGES

Op Amp	$V_{out(off)}$	$V_{in(off)}$	I_{max}
1			
2			
3			

TABLE 42-4. SLEW RATE, POWER BANDWIDTH, AND MPP VALUE

Op Amp	S_R	f_{max}	MPP
1			
2			
3			

TABLE 42-5. TROUBLESHOOTING

Trouble	DC Output Voltage	AC Output Voltage
No +15 V		
No −15 V		
Pin 2 shorted to GND		

TABLE 42-6. CRITICAL THINKING

$R_2 =$ _____

$A =$ _____

Questions for Experiment 42

1. The calculated dc voltages in Table 42-1 are approximately: ( )
   (a) 1 mV; (b) 5.6 mV; (c) 12.3 mV; (d) 17.6 mV.
2. The input bias current of Table 42-2 is closest to: ( )
   (a) 1 nA; (b) 80 nA; (c) 2 mA; (d) 25 mA.
3. The short-circuit currents of Table 42-3 are closest to: ( )
   (a) 1 nA; (b) 80 nA; (c) 2 mA; (d) 25 mA.
4. When the input frequency was much higher than the f_{max} of Table 42-4, the output looked: ( )
   (a) sinusoidal; (b) triangular; (c) square; (d) undistorted. ( )
5. The MPP value of Table 42-4 is closest to:
   (a) 5 mV; (b) 15 V; (c) 25 V; (d) 30 V.
6. Explain the meaning of the input offset current and input bias current.

TROUBLESHOOTING

7. Explain the meaning of the input offset voltage.

8. Describe how you measured the slew rate in this experiment.

CRITICAL THINKING

9. What value did you use for R_2? Why?

10. Optional. Instructor's question.

Small- and Large-Signal Output Impedances

Although you can apply an input voltage as needed to null the output, this direct approach to nulling the output can produce unwanted drift or other degrading effects. This is why it is best to use the nulling method suggested on the data sheet. For a 741, the manufacturer recommends using a potentiometer between pins 1 and 5. In this experiment, you will connect a 741C and use a potentiometer to null the output.

Data sheets list the open-loop output impedance $z_{out(OL)}$ of a 741C as 75 Ω. The closed-loop output impedance $z_{out(CL)}$ is usually much lower, typically less than 1 Ω. This typical value applies only to small signals.

When the signal is large, the output impedance increases. We will define the large-signal output impedance by the symbol $z_{out(MPP)}$. This is the output impedance of an op amp when it is producing the maximum peak-to-peak output signal. The value of $z_{out(MPP)}$ is equal to the load resistance that reduces the MPP to half of the unloaded value. Figure 18-6*b* in your textbook shows that a load resistance of slightly more than 200 Ω reduces the MPP from 27 V (unloaded MPP) to 13.5 V. In this experiment, you will measure the approximate value of $z_{out(MPP)}$.

Required Reading

Chapter 18 (Secs. 18-1 to 18-4) of *Electronic Principles*, 6th ed.

Equipment

1   audio generator
2   power supplies: ±15 V
1   op amp: 741C
8   ½-W resistors: 10 Ω, 100 Ω, 1 kΩ, three 10 kΩ, two 100 kΩ
1   capacitor: 4.7 μF
2   potentiometer: 1 kΩ, 10 kΩ
1   VOM (analog or digital)
1   oscilloscope

Procedure

1. Assume $I_{in(bias)} = 80$ nA, $I_{in(off)} = 20$ nA, and $V_{in(off)} = 2$ mV in Fig. 43-1. With Eqs. (18-7) to (18-10) in your textbook, calculate and record the error voltages listed in Table 43-1.
2. Build the circuit of Fig. 43-1. Measure and record the output error voltage V_{err} in Table 43-1.
3. Note that the input error voltages are superimposed at the input terminals and cannot be measured separately. This is why they are listed as "DNA" (does not apply) in Table 43-1.
4. A circuit like Fig. 43-1 is not compensated for input bias current. When $I_{in(bias)} = 80$ nA, $V_{1err} \approx 0.8$ mV. This is for the typical value of input bias current for a 741C. In the worst case, $I_{in(bias)} = 500$ nA and

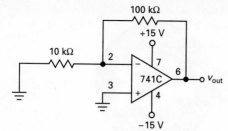

Figure 43-1

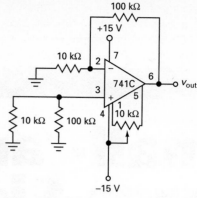

Figure 43-3

$V_{1err} \approx 5$ mV. When V_{1err} is large like this, a designer may use a compensating resistor on the other side of the op amp to reduce the effect of input bias current.

5. Assume $I_{in(bias)} = 80$ nA, $I_{in(off)} = 20$ nA, and $V_{in(off)} = 2$ mV in Fig. 43-2. Calculate and record the error voltages listed in Table 43-2.

6. Build the circuit of Fig. 43-2. Measure and record the output error voltage V_{err} in Table 43-2.

7. A circuit like Fig. 43-2 is compensated for the input bias current. Ideally, $V_{1err} = 0$, leaving only V_{2err} and V_{3err}. The best way to eliminate the effect of these remaining input error voltages is by using the nulling-potentiometer method shown in Fig. 43-3.

8. Connect the circuit of Fig. 43-3.

9. With a DMM, adjust the potentiometer to get an output voltage as close to zero as possible.

10. Incidentally, the parallel equivalent resistance of 10 kΩ and 100 kΩ is 9.09 kΩ. Therefore, we can use the nearest standard value of 9.1 kΩ instead of two separate resistors to compensate for input bias current.

11. After the circuit has been compensated and nulled, we can ac-couple a signal into the circuit to measure small-signal and large-signal output impedances.

12. Connect an audio generator through a 4.7-μF coupling capacitor to the noninverting input as shown in Fig. 43-4.

13. Set the generator frequency to 1 kHz.

14. Measure the output voltage with an oscilloscope. Adjust the generator amplitude to get an output sine wave with a peak-to-peak value of 1 V.

15. Connect a load resistance of 10 kΩ between pin 6 and ground. Measure and record the output voltage in Table 43-3.

16. Repeat Step 15 for the other load resistances shown in Table 43-3. Because of the heavy negative feedback, the closed-loop output impedance will be very small and the load voltage should remain at approximately 1 V pp for all load resistances. In other words, the op amp should act like a stiff voltage source.

17. Disconnect the load resistor and increase the input signal until you get the maximum unclipped output voltage MPP. This should be 27 to 28 V pp. Make sure that the signal is not quite clipping by backing off slightly from the clipped level. Record the value here:

Unloaded MPP = _____ .

18. Connect a load resistance of 10 kΩ between pin 6 and ground. If you see any clipping, reduce the generator amplitude slightly to get the maximum unclipped output. Measure and record the peak-to-peak output voltage in Table 43-4.

19. Repeat Step 18 for the other load resistances shown in Table 43-4. After each load resistor is connected, you

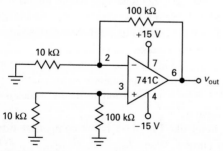

Figure 43-2

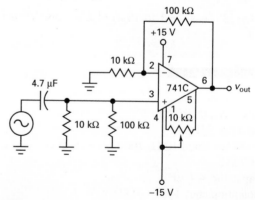

Figure 43-4

will have to reduce the generator amplitude as needed to prevent output clipping. Adjust the level as needed to get the maximum unclipped output voltage.

20. Notice how the MPP decreased in Step 19 as the load resistance decreased. The large-signal output impedance equals the load resistance when the loaded MPP is half of the unloaded MPP.

21. Replace the load resistor by a 1-kΩ potentiometer connected as a variable resistance. Vary the load resistance until the load voltage is half of the unloaded MPP. Disconnect the potentiometer and measure its resistance. Record its value here:

$z_{out(MPP)} =$ _____ .

22. The foregoing value of output impedance applies to large signals only. The large-signal output impedance is always more than the small-signal open-loop output impedance z_{out} shown on data sheets.

Data for Experiment 43

TABLE 43-1. ERROR VOLTAGES WITHOUT COMPENSATION

	Calculated	Measured
V_{1err}		DNA
V_{2err}		DNA
V_{3err}		DNA
V_{error}		

TABLE 43-2. ERROR VOLTAGES WITH COMPENSATION

	Calculated	Measured
V_{1err}		DNA
V_{2err}		DNA
V_{3err}		DNA
V_{error}		

TABLE 43-3. SMALL-SIGNAL OUTPUT IMPEDANCE

R_L	10 kΩ	1 kΩ	100 Ω	10 Ω
V_L				

TABLE 43-4. LARGE-SIGNAL OUTPUT IMPEDANCE

R_L	10 kΩ	1 kΩ	100 Ω	10 Ω
V_L				

Questions for Experiment 43

1. In Fig. 43-1, the input error voltage caused by input bias current is typically:
   (a) 0.1;   (b) 0.8 mV;   (c) 2 mV;   (d) 6 mV.   ( )
2. The resistors connected to pin 3 of Fig. 43-2 compensate for the input:
   (a) offset current;   (b) bias current;   (c) offset voltage;   (d) noise.   ( )
3. The 10-kΩ potentiometer of Fig. 43-3 is used to:
   (a) set A_{CL};   (b) zero the output;   (c) improve ac response;   (d) add feedback.   ( )
4. The 4.7-μF capacitor prevents dc current from flowing through the:
   (a) generator resistance;   (b) 910 Ω;   (c) op-amp noninverting input;   (d) 1 kΩ.   ( )
5. Connecting a small load resistance:
   (a) reduces MPP;   (b) increases MPP;   (c) increases gain;   (d) has no effect.   ( )
6. The output impedance of an op amp increases when the output signal is:
   (a) small;   (b) a dc signal;   (c) large;   (d) zero.   ( )

7. The closed-loop voltage gain of Fig. 43-4 is closest to: ( )
   (a) 1;   **(b)** 10;   **(c)** 11;   **(d)** 100.

8. The data of Table 43-3 implies that the small-signal output impedance is very: ( )
   (a) small;   **(b)** distorted;   **(c)** large;   **(d)** uncompensated.

9. Write a brief explanation of why Step 21 gave you the value of the large-signal output impedance.

10. Optional. Instructor's question.

Summing Amplifiers

The summing amplifier is an inverting amplifier with two or more input channels. Each channel has its own voltage gain given by the ratio of the feedback resistance to the channel resistance. In this experiment, you will connect a summing amplifier and verify that its output voltage is the sum of the input voltages.

Required Reading

Chapter 18 (Sec. 18-5) of *Electronic Principles*, 6th ed.

Equipment

1 audio generator
2 power supplies: ±15 V
1 VOM (analog or digital multimeter)
1 op amp: 741C
3 ½-W resistors: three 10 kΩ
2 switches: SPST
1 oscilloscope

Procedure

1. In the summing amplifier of Fig. 44-1, the source signal is 1 V pp and 1 kHz. Calculate the voltage gain for each channel and record in Table 44-1. Then, calculate and record the peak-to-peak output voltage for the switch positions shown in the table.
2. Connect the circuit of Fig. 44-1. Use channel 1 of the oscilloscope to look at v_1 or v_2. Use channel 2 to look at v_{out}. Measure and record the output voltage for the switch positions shown in Table 44-2. *Note:*

Because the generator has a typical output impedance of 600 Ω, there will be a small loading effect. For this reason, you may need to readjust slightly the generator amplitude to ensure inputs of 1 V pp on different switch positions. In other words, verify that v_1 and v_2 are the values listed in Table 44-2 before measuring v_{out}.

3. In Fig. 44-1, assume that R_1 is replaced by a 22-kΩ resistor. If the switches are closed and both inputs are 1 V pp, what is the output voltage? Record your answer here:

_____ .

4. Replace R_1 by a 33-kΩ resistor. With both switches closed, adjust the generator amplitude so that $v_1 = v_2$ = 1 V pp. Measure and record the output voltage here:

_____ .

5. In Fig. 44-1, assume that R_F = 27 kΩ. If the switches are closed and both inputs are 1 V pp, what is the output voltage? Record your answer here:

_____ .

6. Assume that $R_1 = R_2 = 10$ kΩ and $R_F = 27$ kΩ in Fig. 44-1. With the switches closed, calculate the output voltage for each input shown in Table 44-3.
7. Connect the circuit with $R_1 = R_2 = 10$ kΩ and $R_F = 27$ kΩ. With both switches closed, adjust the generator

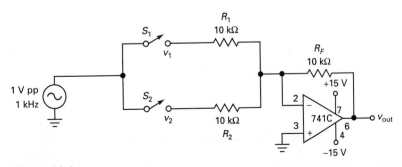

Figure 44-1

amplitude so that v_1 equals the values listed in Table 44-3. Measure and record the output voltage.

COMPUTER (OPTIONAL)

8. Repeat Steps 1 to 7 using EWB or an equivalent circuit simulator. Do not record any new values. But make sure that you get reasonable agreement between the EWB measurements and the values recorded earlier.
9. If you are using the CD-ROM version of this book, click on the Assignments menu and select Chap. 18.

ADDITIONAL WORK (OPTIONAL)

10. The circuit of Fig. 44-2 contains positive and negative clippers. With a sinusoidal source, two opposite-polarity half-wave signals drive the summing amplifier. Answer the following questions.
    a. If S_1 is closed and S_2 is open, what does the output of the summing amplifier look like?

    Answer = _____ .
    b. With S_1 and S_2 closed, what does the upper 100-kΩ variable resistor do?

    Answer = _____ .
    c. With S_1 and S_2 closed, what does the 10-kΩ variable resistor do?

    Answer = _____ .
    d. What is the maximum voltage gain on channel 1?

    Answer = _____ .
    e. What is the minimum voltage gain on channel 2?

    Answer = _____ .
11. Construct the circuit. Be sure to connect the output of the clippers to the points A and B on the switches.
12. Adjust the audio generator to 20 V pp and 1 kHz. This will produce half-wave signals at points A and B. If you cannot get this much output, adjust the generator amplitude to its maximum.

13. Close S_1 and open S_2. Adjust the 10-kΩ variable resistance to maximum. With the oscilloscope, look at point A while changing the upper 100-kΩ variable resistor over its entire range. Describe the output signal:

14. Adjust the upper 100-kΩ variable resistor to get an output voltage with a positive peak of 5 V.
15. Open S_1 and close S_2. With the oscilloscope, look at point B while changing the lower 100-kΩ variable resistor over its entire range. Describe the output signal:

16. Adjust the lower 100-kΩ variable resistor to get an output voltage with a negative peak of −5 V.
17. Close S_1 and S_2. Describe what you see:

18. Vary the 10-kΩ feedback resistor through its entire range. Describe the effect it has on the output signal:

19. Vary the upper and lower 100-kΩ potentiometers and observe the effect these have on each half cycle of the output. Describe what you see:

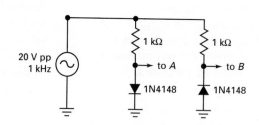

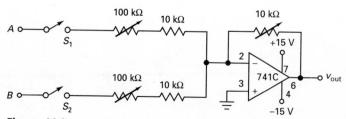

Figure 44-2

194

Data for Experiment 44

TABLE 44-1. CALCULATIONS: $A_1 =$ _____ AND $A_2 =$ _____

S_1	S_2	v_1	v_2	v_{out}
Open	Open	0	0	
Open	Closed	0	1 V pp	
Closed	Open	1 V pp	0	
Closed	Closed	1 V pp	1 V pp	

TABLE 44-2. MEASUREMENTS: $A_1 =$ _____ AND $A_2 =$ _____

S_1	S_2	v_1	v_2	v_{out}
Open	Open	0	0	
Open	Closed	0	1 V pp	
Closed	Open	1 V pp	0	
Closed	Closed	1 V pp	1 V pp	

TABLE 44-3. S_1 AND S_2 CLOSED

v_1 or v_2:	0.5 V pp	1 V pp	1.5 V pp	2 V pp
Calculated v_{out}:				
Measured v_{out}:				

Questions for Experiment 44

1. In Fig. 44-1, the voltage gain of each channel is: ( )
   (a) 0; (b) 1; (c) 2; (d) 10.
2. If each input voltage is 1 V pp in Fig. 44-1, the peak-to-peak output is: ( )
   (a) 0; (b) 1; (c) 2; (d) 10.
3. If R_1 is changed to 33 kΩ in Fig. 44-1, the peak-to-peak output is: ( )
   (a) 0; (b) 1; (c) 1.3; (d) 5.4.
4. If R_F is changed to 27 kΩ in Fig. 44-1, the peak-to-peak output is: ( )
   (a) 0; (b) 1; (c) 1.3; (d) 5.4.
5. If S_1 is open and S_2 is closed in Fig. 44-1, the peak-to-peak output is: ( )
   (a) 0; (b) 1; (c) 2; (d) 10.
6. A summing amplifier has $v_1 = 1.5$ V dc, $v_2 = -2$ V, and unity voltage gain on ( )
   each channel. Because of the phase inversion, the output voltage is:
   (a) 0.5 V; (b) -0.5 V; (c) 3.5 V; (d) -3.5 V.
7. Many factors may contribute to the discrepancies between the calculated values of Table
   44-1 and the measured values of Table 44-2. But there is one factor that causes more error
   than any other. State what it is and why it produces the errors:

8. Suppose you have the following signals: the output of a microphone, the output of a CD player, and the audio output of a video camera. You want to mix these signals and record the combined output on a tape recorder. Describe one way to do it:

9. Optional. Instructor's question.

10. Optional. Instructor's question.

VCVS Feedback

There are four basic types of negative feedback, depending on which input is used and which output quantity is sampled. VCVS feedback results in an almost perfect voltage amplifier, one with high input impedance, low output impedance, and stable voltage gain. The negative feedback also reduces nonlinear distortion and output offset voltage.

In this experiment you will work with VCVS feedback. First, you will see how accurate the formula for closed-loop voltage gain is. Second, you will see how stable the voltage gain is for different op amps. Third, you will calculate and measure output offset voltages for different feedback resistors. Also included are troubleshooting, design, and computer options.

Required Reading

Chapter 19 (Secs. 19-1 to 19-2) of *Electronic Principles*, 6th ed.

Equipment

1 audio generator
2 power supplies: ±15 V
9 ½-W resistors: two 1 kΩ, two 10 kΩ, 22 kΩ, 33 kΩ, 47 kΩ, 68 kΩ, 100 kΩ
3 op amps: 741C
2 capacitors: 0.47 μF
1 VOM (analog or digital multimeter)
1 oscilloscope

Procedure

VOLTAGE AMPLIFIER

1. In Fig. 45-1, assume that R_1 equals 10 kΩ. Calculate the closed-loop voltage gain. Record A_{CL} in Table 45-1.
2. Repeat Step 1 for the other values of R_1 shown in Table 45-1.
3. Connect the voltage amplifier of Fig. 45-1 with R_1 equal to 10 kΩ. Set the audio generator to 1 kHz at 100 mV pp. Measure v_{out}. Calculate the closed-loop voltage gain with $A_{CL} = v_{out}/v_{in}$. Record this as the measured A_{CL}.
4. Repeat Step 3 for the other values of R_1 listed in Table 45-1.

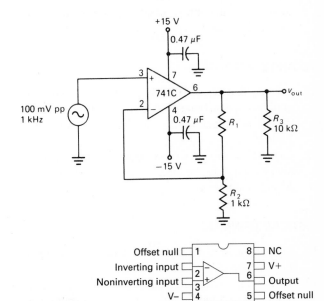

Figure 45-1

STABLE VOLTAGE GAIN

5. Assume that R_1 is 33 kΩ in Fig. 45-1. Calculate and record the closed-loop voltage gain (Table 45-2).
6. Connect the circuit with R_1 equal to 33 kΩ. Measure v_{out} and calculate A_{CL}. Record this measured value in Table 45-2.
7. Repeat Steps 5 and 6 for the other 741Cs.

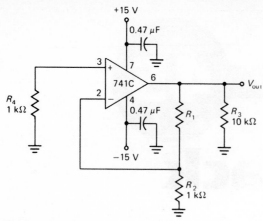

Figure 45-2

OUTPUT OFFSET VOLTAGE

8. Differences in the V_{BE} values and the input base currents imply there is a dc input offset voltage in Fig. 45-2. Assume that the total input offset voltage is 2 mV. Calculate and record the output offset voltage for each value of R_1 listed in Table 45-3.

9. Connect the circuit. Measure and record the output offset voltage for each value of R_1. (Even though your measured values may differ considerably from calculated values, the output offset voltage should increase with an increase in R_1.)

TROUBLESHOOTING

10. Assume that R_1 is 100 kΩ in Fig. 45-1. For each trouble listed in Table 45-4, estimate the dc and peak-to-peak ac output voltage. Record your estimates in Table 45-4.

11. Connect the circuit with R_1 equal to 100 kΩ. Insert each trouble into the circuit. Measure and record the dc and ac output voltages.

CRITICAL THINKING

12. Select a value of R_1 in Fig. 45-1 to set up a closed-loop voltage gain of 40.

13. Connect the circuit with your design value of R_1. Measure the closed-loop voltage gain. Record your R_1 and A_{CL} in Table 45-5.

COMPUTER (OPTIONAL)

14. Repeat Steps 1 to 13 using EWB or an equivalent circuit simulator. Do not record any new values. But make sure that you get reasonable agreement between the EWB measurements and the values recorded earlier.

15. If you are using the CD-ROM version of this book, click on the Assignments menu and select Chap. 19.

ADDITIONAL WORK (OPTIONAL)

16. A 741C has a typical short-circuit output current of ±25 mA, or a peak-to-peak swing of 50 mA. Estimate the MPP for an 8-Ω speaker by multiplying 50 mA by 8 Ω. Record your answer here:

_____ .

17. Calculate the speaker power using the MPP of the previous step.

18. Do you think that a voltage follower using a 741C can drive a small 8-Ω speaker with enough power to produce an audible sound? You are about to find out.

19. Connect the circuit of Fig. 45-3.

20. Use channel 1 to measure the input voltage and channel 2 to measure the output voltage. Use a sensitivity of 0.1 V/DIV on each channel.

21. Adjust the peak-to-peak input voltage to 300 mV pp and 1 kHz. You should hear a 1-kHz tone. Vary the frequency, and the tone will change.

22. If the output waveform is clipped or distorted, reduce the input slightly until the clipping disappears.

23. Compare the input and output voltage waveforms. The two waveforms should be approximately the same since the circuit is a voltage follower. Since the load impedance is only 8 Ω, this means that a voltage follower is a stiff voltage source. The limitation on output power is the short-circuit current of a 741C, which is only ±25 mA. Do you have any ideas on how to get more speaker power?

24. What did you learn in Steps 16 to 23? Record your summary here:

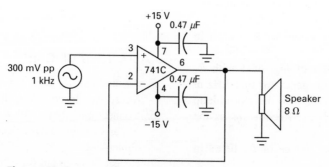

Figure 45-3

Data for Experiment 45

TABLE 45-1. CLOSED-LOOP VOLTAGE GAIN

R_1	Calculated A_{CL}	Measured A_{CL}
10 kΩ		
22 kΩ		
47 kΩ		
68 kΩ		
100 kΩ		

TABLE 45-2. STABLE VOLTAGE GAIN

Op Amp	Calculated A_{CL}	Measured A_{CL}
1		
2		
3		

TABLE 45-3. CLOSED-LOOP OUTPUT OFFSET VOLTAGE

R_1	Calculated $V_{oo(CL)}$	Measured $V_{oo(CL)}$
10 kΩ		
22 kΩ		
47 kΩ		
68 kΩ		
100 kΩ		

TABLE 45-4. TROUBLESHOOTING

| Trouble | Estimated | | Measured | |
	V_{out}	v_{out}	V_{out}	v_{out}
R_1 short				
R_1 open				
R_2 short				
R_2 open				

TABLE 45-5. CRITICAL THINKING

$R_1 =$

$A_{CL} =$

Questions for Experiment 45

1. The calculated and measured A_{CL} of Table 45-1 were: ( )
   (a) extremely large;   (b) very small;   (c) close in value;
   (d) unpredictable.

2. The measured A_{CL} of Table 45-2 for all three 741Cs was: ( )
   (a) extremely large;   (b) very small;   (c) almost constant;   (d) quite
   variable.

3. When R increases in Table 45-3, the closed-loop voltage gain increases and the   ( )
   output offset voltage:
   (a) decreases;   (b) increases;   (c) stays the same;   (d) none of the
   foregoing.

4. The closed-loop voltage gain of an amplifier with noninverting voltage feedback   ( )
   is as stable as the:
   (a) supply voltage;   (b) gain of the 741C;   (c) load resistor;
   (d) feedback resistors.

5. If the input bias current is 80 nA in Fig. 45-2, the dc voltage across R_4 is: ( )
   (a) 80 μV;   (b) 800 μV;   (c) 2 mV;   (d) -15 V.

6. What is the ac voltage at the inverting input of Fig. 45-1? Why?

TROUBLESHOOTING

7. When R_1 is open or R_2 is shorted in Fig. 45-1, you get a clipped output with a peak-to-peak
   value around 28 V. Explain why this happens.

8. When R_1 is shorted or R_2 is open in Fig. 45-1, what does the closed-loop voltage gain equal?
   What is the name for this kind of circuit?

CRITICAL THINKING

9. You are designing a voltage amplifier like Fig. 45-1. If you want to get a voltage gain
   accurate to within 2 percent, what would you specify in your design?

10. Optional. Instructor's question.

Negative Feedback

Always remember that there are four distinct types of negative feedback. Each type has different characteristics. VCVS feedback results in a voltage amplifier. VCIS feedback leads to a voltage-to-current converter. ICVS feedback results in a current-to-voltage converter. ICIS feedback leads to a current amplifier.

All four types of negative feedback reduce nonlinear distortion and output offset voltage. The noninverting types increase the input impedance, while the inverting types decrease it. The voltage feedback types decrease the output impedance, while the current feedback types increase the output impedance.

In this experiment, you will connect all four types of negative-feedback circuits using dc input and output voltages and currents.

Required Reading

Chapter 19 (Secs. 19-1 to 19-6) of *Electronic Principles,* 6th ed.

Equipment

1 audio generator
2 power supplies: ± 15 V
6 ½-W resistors: two 1 kΩ, 2 kΩ, two 10 kΩ, 18 kΩ
1 potentiometer: 1 kΩ
1 op amp: 741C
2 capacitors: 0.47 μF
2 VOMs: If two VOMs are not available, you can run the experiment with only one

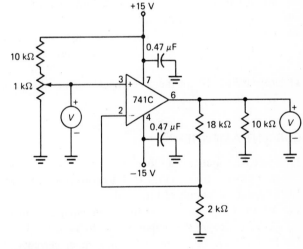

Figure 46-1

Procedure

VOLTAGE AMPLIFIER

1. For each dc input voltage listed in Table 46-1, calculate the dc ouput voltage of Fig. 46-1. Record your answers.
2. Connect the circuit. Use one VOM on the input side and one on the output side. (If you don't have two VOMs, you will have to measure the input voltage first, then the output voltage.)
3. Adjust the potentiometer to get each dc input voltage listed in Table 46-1. Measure and record the output voltage.

VOLTAGE-TO-CURRENT CONVERTER

4. For each dc input voltage in Table 46-2, calculate the dc output current of Fig. 46-2. Record your answers.
5. Connect the circuit of Fig. 46-2. Use one VOM to measure the input voltage and one VOM to measure the output current. (If you have only one VOM to work with, use a short in the place of the output ammeter when measuring the input voltage. When measuring output current, replace the short by the ammeter.)

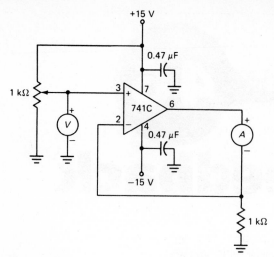

Figure 46-2

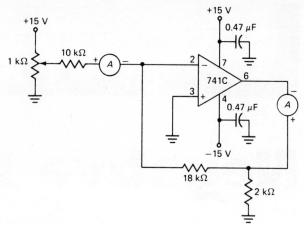

Figure 46-4

6. Adjust the potentiometer to get an input voltage of 1 V. Read the output current and record the value in Table 46-2.

7. Repeat Step 6 for the remaining input voltages listed in Table 46-2.

CURRENT-TO-VOLTAGE CONVERTER

8. For each input current listed in Table 46-3, calculate the output voltage in Fig. 46-3. Record your answers.

9. Connect the circuit of Fig. 46-3.

10. Adjust the potentiometer to get an input current of 1 mA. Read the output voltage and record the value in Table 46-3.

11. Repeat Step 10 for the other input currents shown in Table 46-3.

CURRENT AMPLIFIER

12. For each input current listed in Table 46-4, calculate the output current in Fig. 46-4. Record your answers.

13. Connect the circuit of Fig. 46-4.

14. Adjust the potentiometer to get an input current of 0.1 mA. Record the output current in Table 46-4.

15. Repeat Step 14 for the remaining input currents of Table 46-4.

TROUBLESHOOTING

16. Ask the instructor to insert a trouble into any circuit he or she wishes.

17. Locate and repair the trouble. Record each trouble in Table 46-5.

18. Repeat Steps 16 and 17 as often as indicated by the instructor.

CRITICAL THINKING

19. The circuit of Fig. 46-2 has a transconductance of 100 μS. Redesign the circuit so that it has a g_m of 500 μS.

20. Connect the redesigned circuit. Measure the output current for each voltage listed in Table 46-6.

COMPUTER (OPTIONAL)

21. Repeat Steps 1 to 20 using EWB or an equivalent circuit simulator. Do not record any new values. But make sure that you get reasonable agreement between the EWB measurements and the values recorded earlier.

22. If you are using the CD-ROM version of this book, click on the Assignments menu and select Chap. 19.

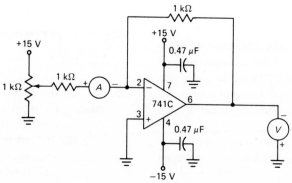

Figure 46-3

Data for Experiment 46

TABLE 46-1. VCVS FEEDBACK

V_{in}	Calculated V_{out}	Measured V_{out}
0.1 V		
0.2 V		
0.3 V		
0.4 V		
0.6 V		
0.8 V		
1 V		

TABLE 46-2. VCIS FEEDBACK

V_{in}	Calculated I_{out}	Measured I_{out}
1 V		
2 V		
3 V		
4 V		
6 V		
8 V		
10 V		

TABLE 46-3. ICVS FEEDBACK

I_{in}	Calculated V_{out}	Measured V_{out}
1 mA		
2 mA		
3 mA		
4 mA		
6 mA		
8 mA		
10 mA		

TABLE 46-4. ICIS FEEDBACK

I_{in}	Calculated I_{out}	Measured I_{out}
0.1 mA		
0.2 mA		
0.3 mA		
0.4 mA		
0.6 mA		
0.8 mA		
1 mA		

TABLE 46-5. TROUBLESHOOTING

Trouble	Description
1	
2	
3	

TABLE 46-6. CRITICAL THINKING

V_{in}	I_{out}
1 V	
2 V	
3 V	
4 V	
6 V	
8 V	
10 V	

Questions for Experiment 46

1. The voltage gain of Table 46-1 is closest to: ( )
   (a) 1; (b) 5; (c) 10; (d) 20.
2. The transconductance of Table 46-2 is approximately: ( )
   (a) 100 μS; (b) 300 μS; (c) 750 μS; (d) 1000 μS.
3. The transresistance of Table 46-3 is approximately: ( )
   (a) 100 Ω; (b) 1 kΩ; (c) 10 kΩ; (d) 100 kΩ.
4. The current gain of Table 46-4 is closest to: ( )
   (a) 1; (b) 10; (c) 100; (d) 1000.
5. The stability or accuracy of any of the feedback circuits in this experiment depends ( )
   primarily on the:
   (a) supply voltage; (b) 741C; (c) VOM; (d) tolerance of feed-back resistors.
6. What did you learn from this experiment? List at least two ideas that seem important to you.

204

TROUBLESHOOTING

7. Somebody mistakenly uses a 10-kΩ resistor for the feedback resistor of Fig. 46-2. How will this affect the circuit performance?

8. The negative supply voltage of Fig. 46-3 is not connected to the op amp. What are the symptoms of this trouble?

CRITICAL THINKING

9. You are designing an electronic VOM. Which of the basic feedback circuits would you use to measure voltage? Which would you use for measuring current?

10. Optional. Instructor's question.

Gain-Bandwidth Product

Whenever you work with an op amp, remember that the gain-bandwidth product is a constant. This means the product of closed-loop voltage gain and bandwidth equals the unity-gain frequency of the op amp. Stated another way, it means you can trade off voltage gain for bandwidth. For instance, if you reduce the voltage gain by a factor of 2, you will double the bandwidth.

In this experiment, you will calculate and measure the bandwidth for different voltage gains. This will confirm that the gain-bandwidth product is a constant.

Required Reading

Chapter 19 (Sec. 19-7) of *Electronic Principles*, 6th ed.

Equipment

1 sine/square generator
2 power supplies: ±15 V
6 $\frac{1}{2}$-W resistors: 4.7 kΩ, 6.8 kΩ, 10 kΩ, 22 kΩ, 33 kΩ, 47 kΩ
1 op amp: 741C
2 capacitors: 0.47 μF
1 VOM (analog or digital multimeter)
1 oscilloscope
1 frequency counter

Figure 47-1

Procedure

CALCULATING VOLTAGE GAIN AND BANDWIDTH

1. For each R listed in Table 47-1, calculate the closed-loop voltage gain of Fig. 47-1. Record all answers.
2. The typical gain-bandwidth product (same as f_{unity}) of a 741C is 1 MHz. Calculate and record the closed-loop cutoff frequency for each R listed in Table 47-1.
3. Connect the circuit with R equal to 4.7 kΩ. Look at the output signal with an oscilloscope. With the input frequency at 100 Hz, adjust the signal level to get an output of 5 V pp.
4. Measure the peak-to-peak input voltage. Calculate and record A_{CL} as a measured quantity (Table 47-1).

5. Measure and record the upper cutoff frequency.
6. Repeat Steps 3 to 5 for the other values of R in Table 47-1.

MEASURING RISETIME TO GET BANDWIDTH

7. Connect the circuit of Fig. 47-1 with an R of 4.7 kΩ and a square-wave generator instead of a sine-wave generator.
8. With the frequency around 5 kHz, adjust the signal level to get an output voltage of 5 V pp.
9. Measure the risetime and record in Table 47-2. Calculate and record $f_{2(CL)}$.

10. Repeat Steps 7 to 9 for the other values of R. (Note: You will have to use an input frequency less than 5 kHz as the value of R increases. Reduce the frequency as needed to get an accurate risetime measurement.)

TROUBLESHOOTING

11. Estimate the risetime in Fig. 47-1 for each trouble listed in Table 47-3. Record your answers.
12. Insert each trouble. Measure and record the risetime.

CRITICAL THINKING

13. Select a value of R in Fig. 47-1 to get a bandwidth of 35 kHz (Use a 741C.)

14. Connect the circuit of Fig. 47-1 with your value of R. Measure the voltage gain and risetime. Calculate the bandwidth. Record all quantities listed in Table 47-4.

COMPUTER (OPTIONAL)

15. Repeat Steps 1 to 14 using EWB or an equivalent circuit simulator. Do not record any new values. But make sure that you get reasonable agreement between the EWB measurements and the values recorded earlier.
16. If you are using the CD-ROM version of this book, click on the Assignments menu and select Chap. 19.

Data for Experiment 47

TABLE 47-1. GAIN AND CRITICAL FREQUENCY

	Calculated		Measured	
R	A_{CL}	$f_{2(CL)}$	A_{CL}	$f_{2(CL)}$
4.7 kΩ				
6.8 kΩ				
10 kΩ				
22 kΩ				
33 kΩ				
47 kΩ				

TABLE 47-2. RISETIME

R	Measured T_R	Experimental $f_{2(CL)}$
4.7 kΩ		
6.8 kΩ		
10 kΩ		
22 kΩ		
33 kΩ		
47 kΩ		

TABLE 47-3. TROUBLESHOOTING

Trouble	Estimated T_R	Measured T_R
R shorted		
No +15-V supply		
100 Ω open		

TABLE 47-4. CRITICAL THINKING

$R =$	
$A_{CL} =$	
$T_R =$	
$f_{2(CL)} =$	

Questions for Experiment 47

1. The measured data of Table 47-1 indicate that the product of gain and bandwidth is: (   )
(a) 1 MHz;   **(b)** approximately constant;   **(c)** variable;   **(d)** none of the foregoing.

2. The largest value of R in Table 47-2 produces the: ( )
   (a) smallest T_R; (b) largest T_R; (c) smallest voltage gain; (d) none of the foregoing.

3. In Fig. 47-1, an increase in voltage gain leads to a: ( )
   (a) decrease in bandwidth; (b) increase in bandwidth; (c) loss of supply voltage; (d) smaller output voltage.

4. If an op amp has a higher f_{unity}, you can get more bandwidth for a given: ( )
   (a) supply voltage; (b) voltage gain; (c) output voltage; (d) MPP value.

5. To increase the bandwidth of a circuit like Fig. 47-1, you have to: ( )
   (a) decrease the voltage gain; (b) increase the supply voltage;
   (c) decrease the f_{unity}; (d) increase the output voltage.

6. Why is it important to know that the gain-bandwidth product is constant?

TROUBLESHOOTING

7. Suppose one of the bypass capacitors of Fig. 47-1 shorts out. What symptoms will you get?

8. There is no dc or ac output voltage in a circuit like Fig. 47-1. Name three possible causes.

CRITICAL THINKING

9. You are designing an amplifier to have as fast a risetime as possible. Do you want an op amp with a low or a high f_{unity}? Why?

10. Optional. Instructor's question.

Linear IC Amplifiers

Linear op-amp circuits preserve the shape of the input signal. If the input is sinusoidal, the output will be sinusoidal. Two basic voltage amplifiers are possible: the noninverting amplifier and the inverting amplifier. The inverting amplifier consists of a source resistance cascaded with a current-to-voltage converter. As discussed in your textbook, the closed-loop voltage gain equals the ratio of the feedback resistance to the source resistance.

In this experiment you will build and test both types of voltage amplifiers. You will also connect a noninverter/inverter with a single adjustment that allows you to vary the voltage gain.

Required Reading

Chapter 20 (Secs. 20-1 to 20-6) of *Electronic Principles,* 6th ed.

Equipment

- 1 sine/square generator
- 2 power supplies: ±15 V
- 13 ½-W resistors: 100 Ω, two 1 kΩ, 1.1 kΩ, two 6.8 kΩ, 10 kΩ, 47 kΩ, 68 kΩ, 100 kΩ, 220 kΩ, 330 kΩ, 470 kΩ
- 1 potentiometer: 1 kΩ
- 1 op amp: 741C
- 4 capacitors: two 0.47 μF, two 1 μF
- 1 VOM (analog or digital multimeter)
- 1 oscilloscope
- 1 frequency counter

Procedure

SINGLE-SUPPLY NONINVERTING AMPLIFIER

1. Assume a typical f_{unity} of 1 MHz for the 741C of Fig. 48-1. Calculate A_{CL} and $f_{2(CL)}$. Also calculate the input, output, and bypass critical frequencies. Estimate the MPP value. Record all answers in Table 48-1.
2. Connect the circuit. Adjust the audio generator to 100 mV pp at 1 kHz. Measure and record A_{CL}.
3. Measure and record the upper cutoff frequency. (Try both the sine- and the square-wave methods.)
4. Measure and record the lower cutoff frequency.
5. Measure and record the MPP value.

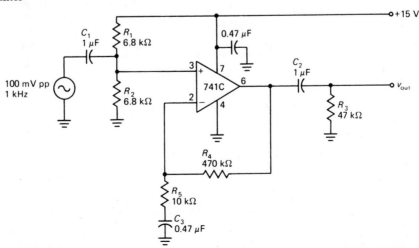

Figure 48-1

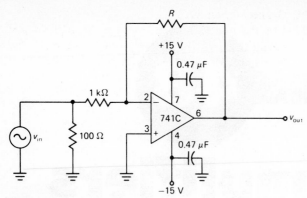

Figure 48-2

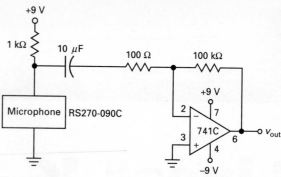

Figure 48-4

INVERTING AMPLIFIER

6. For each R value of Table 48-2, calculate A_{CL} and $f_{2(CL)}$ in Fig. 48-2.
7. Connect the circuit with R equal to 4.7 kΩ. Set the input frequency to 100 Hz. Adjust the signal level to get an output of 5 V pp.
8. Measure v_{in}. Calculate and record A_{CL} as a measured quantity.
9. Measure and record $f_{2(CL)}$.
10. Repeat Steps 7 to 9 for other R values in Table 48-2.

NONINVERTER/INVERTER

11. Calculate the maximum noninverting and inverting voltage gains for the circuit of Fig. 48-3. Record in Table 48-3.
12. Connect the circuit.
13. Look at the output signal with an oscilloscope. Vary the potentiometer and notice what happens.
14. Measure the maximum noninverting and inverting voltage gains. Record the data in Table 48-3.

TROUBLESHOOTING

15. For each trouble listed in Table 48-4, estimate and record the dc voltage at pin 6 (Fig. 48-1).
16. Insert each trouble into the circuit. Measure and record the dc voltage at pin 6.

CRITICAL THINKING

17. Select new values for C_1 and C_3 to get a lower cutoff frequency in Fig. 48-1 that is less than 20 Hz.
18. Connect the circuit. Measure and record the lower cutoff frequency. Record all quantities listed in Table 48-5.

APPLICATION (OPTIONAL)—MICROPHONE PREAMP

19. In Fig. 48-4, the Radio Shack 270-090C is a microphone. It is variable resistance that changes when sound waves strike it. The changing resistance produces an ac voltage, which is an electrical representation of the sound.
20. Connect the circuit of Fig. 48-4.
21. The amplitude of the output signal will depend on how close you are to the microphone when you speak. Look at the output signal of the op amp with an oscilloscope. Set the sensitivity to 20 mV/DIV and the time base to 10 ms/DIV.
22. Watch the output signal while you speak into the microphone. You will see a changing signal when you speak. Experiment with different sensitivities from 10 to 200 mV/DIV as you speak into the microphone. Also try different time bases from 1 to 20 ms/DIV. (What should you say when speaking into the microphone? You might try something like the immortal words of Alexander Graham Bell's first phone call: "Come here, Watson, I need you.")

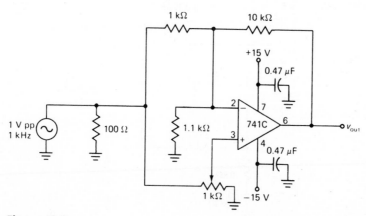

Figure 48-3

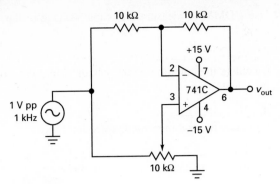

Figure 48-5

COMPUTER (OPTIONAL)

23. Repeat Steps 1 to 18 using EWB or an equivalent circuit simulator. Do not record any new values. But make sure that you get reasonable agreement between the EWB measurements and the values recorded earlier.

24. If you are using the CD-ROM version of this book, click on the Assignments menu and select Chap. 20.

ADDITIONAL WORK (OPTIONAL)—REVERSIBLE AND ADJUSTABLE GAIN

25. The circuit of Fig. 48-5 allows you to vary the voltage gain continuously from −1 to 0, and then from 0 to +1.

26. Connect the circuit.

27. Use the oscilloscope to look at the input signal (channel 1) and the output signal (channel 2). Set both sensitivities to 0.5 V/DIV. Adjust the generator amplitude to get an output signal of 1 V pp.

28. Look at the input and signals while you vary the 10-kΩ potentiometer through its entire range. Describe what the circuit does:

PHASE SHIFTER

29. Connect the circuit of Fig. 48-6.

30. Use channel 1 of the oscilloscope for the input signal and channel 2 for the output signal. Set both sensitivities to 0.5 V/DIV. Adjust the generator amplitude to get an output voltage of 1 V pp and a frequency of 1 kHz.

31. Look at the input and output signals while varying the resistor through its range. Describe what the circuit does:

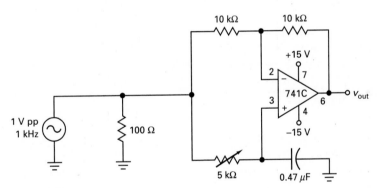

Figure 48-6

WHEATSTONE BRIDGE AND DIFFERENTIAL AMPLIFIER

32. Connect the circuit of Fig. 48-7.

33. Use a DMM to measure and record the differential voltage out of the bridge:

$v_{in} = $ _____ .

34. Use a DMM to measure the output voltage of the op amp:

$v_{out} = $ _____ .

35. Calculate the differential voltage gain:

$A = $ _____ .

36. Place a short between points A and B. In this case, the input is a common-mode signal because $v_1 = v_2$. Use the DMM to measure v_1 with respect to ground:

$v_{in(CM)} = $ _____ .

37. Use a DMM to measure the output voltage of the op amp:

$v_{out(CM)} = $ _____ .

38. Calculate the common-mode voltage gain and record its value:

$A_{CM} = $ _____ .

39. Calculate the common-mode rejection ratio:

CMRR $= $ _____ .

40. What did you learn in Steps 33 to 39?

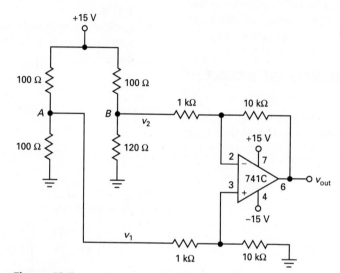

Figure 48-7

Data for Experiment 48

TABLE 48-1. NONINVERTING AMPLIFIER

Calculated

$A_{CL} =$

$f_{2(CL)} =$

$f_{in} =$

$f_{out} =$

$f_{BY} =$

MPP =

Measured

$A_{CL} =$

$f_{2(CL)} =$

$f_{1(CL)} =$

MPP =

TABLE 48-2

| | Calculated | | Measured | |
R	A_{CL}	$f_{2(CL)}$	A_{CL}	$f_{2(CL)}$
47 kΩ				
68 kΩ				
100 kΩ				
220 kΩ				
330 kΩ				
470 kΩ				

TABLE 48-3. NONINVERTER/INVERTER

Calculated

$A_{non} =$

$A_{inv} =$

Measured

$A_{non} =$

$A_{inv} =$

TABLE 48-4. TROUBLESHOOTING

Trouble	DC Voltage at Pin 6
R_1 open	
R_1 short	
R_2 open	
R_2 short	
C_1 open	

TABLE 48-5. CRITICAL THINKING

$C_1 =$	
$C_3 =$	
$f_{1(CL)} =$	

Questions for Experiment 48

1. The MPP value in Table 48-1 is closest to:　　　　　　　　　　　( )
   (a) 1 V;　　**(b)** 7.5 V;　　**(c)** 12.5 V;　　**(d)** 20 V.
2. The product of voltage gain and bandwidth in Table 48-2 is:　　( )
   (a) approximately constant;　　**(b)** small;　　**(c)** 100;　　**(d)** 20 kHz.
3. The noninverter/inverter of Fig. 48-3 has a noninverting voltage gain of　( )
   approximately:
   (a) 1;　　**(b)** 10;　　**(c)** 100;　　**(d)** 1000.
4. The bypass capacitor of Fig. 48-1 sets up a cutoff frequency of approximately:　( )
   (a) 3.39 Hz;　　**(b)** 33.9 Hz;　　**(c)** 46.8 Hz;　　**(d)** 63 Hz.
5. With the inverter of Fig. 48-2, you can increase the bandwidth by:　　( )
   (a) decreasing the supply voltage;　　**(b)** decreasing the voltage gain;
   (c) increasing the value of R;　　**(d)** eliminating the bypass capacitors.
6. Explain how the inverter of Fig. 48-2 works.

TROUBLESHOOTING

7. What happens to the dc voltage at pin 6 when R_1 is shorted? Why?

8. Suppose the bypass capacitor C_3 opens in Fig. 48-1. What kind of dc and ac symptoms will you get at the output of the op amp?

CRITICAL THINKING

9. Explain why you selected the C_1 and C_3 in your design.

10. Optional. Instructor's question.

Current Boosters and Controlled Current Sources

For a given input voltage, the load current of a VCVS amplifier depends on the load resistance. The smaller the load resistance, the larger the current. A typical op amp like the 741 cannot drive very small load resistances because its maximum output current is 25 mA. When a load requires more current than this, you can use a current booster on the output of the op amp. In this experiment, you will build and test a voltage follower with a current booster.

The voltage-controlled current source uses VCIS negative feedback. Because of this, a given input voltage sets up a constant load current. With a VCIS amplifier, you can change load resistance without changing the load current. In this experiment, you will also build and test a voltage-controlled current source.

Required Reading

Chapter 20 (Secs. 20-7 and 20-8) of *Electronic Principles,* 6th ed.

Equipment

1 audio generator
2 power supplies: ±15 V
9 ½-W resistors: 10 Ω, 33 Ω, 100 Ω, 330 Ω, three 1 kΩ, 1.8 kΩ, 2.2 kΩ
1 potentiometer: 1 kΩ
3 transistors: 2N3904, 2N3906, 2N3055
2 op amps: 741C
4 capacitors: 0.47 μF
1 VOM (analog or digital multimeter)
1 oscilloscope

Procedure

VOLTAGE FOLLOWER WITHOUT CURRENT BOOSTER

1. Connect the circuit of Fig. 49-1 with $R_L = 1$ kΩ.

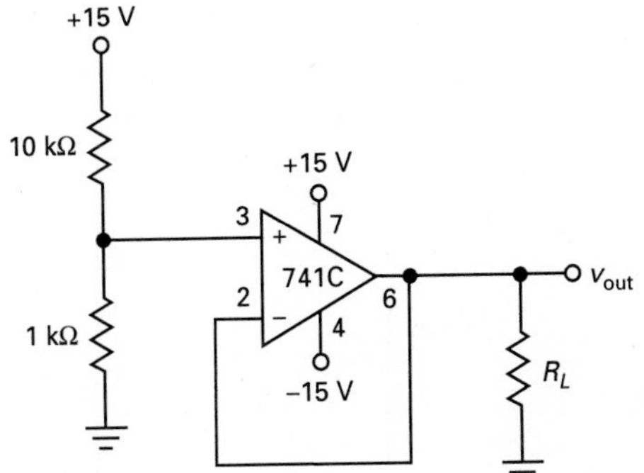

Figure 49-1

2. Measure the input voltage to pin 3 and record the value in Table 49-1.
3. Measure and record the load voltage and load current for each of the load resistances shown in Table 49-1.
4. State your observations about the load current for smaller resistances:

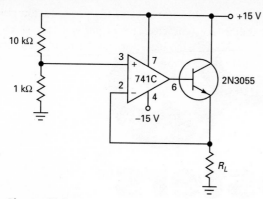

Figure 49-2

VOLTAGE FOLLOWER WITH CURRENT BOOSTER

5. Add a current booster to the voltage follower, as shown in Fig. 49-2.
6. Repeat Steps 2 and 3 for Table 49-2.
7. What have you learned about an op amp used without a current booster? With a current booster?

VOLTAGE-CONTROLLED CURRENT SOURCE

8. In Fig. 49-3, assume that R_L is 100 Ω and calculate the load current for each input voltage listed in Table 49-3.
9. Assume that R_L is 2.2 kΩ and calculate the load current for each input voltage in Table 49-4.

10. Connect the circuit with R_L equal to 100 Ω.
11. Adjust the input voltage to each value in Table 49-3. Measure and record the load current.
12. Change R_L to 2.2 kΩ. Measure the load current for each input voltage of Table 49-4.

TROUBLESHOOTING

13. In Fig. 49-3, assume an R_L of 100 Ω and a v_{in} of +5 V. Estimate and record the load current for each trouble of Table 49-5.
14. Insert each trouble into the circuit. Measure and record the load current.

CRITICAL THINKING

15. Redesign the circuit of Fig. 49-3 to get a load current of approximately 2.25 mA when v_{in} is +5 V.
16. Connect the circuit with your design values. Measure the load current with a v_{in} of +5 V. Record quantities listed in Table 49-6.

COMPUTER (OPTIONAL)

17. Repeat Steps 1 to 16 using EWB or an equivalent circuit simulator. Do not record any new values. But make sure that you get reasonable agreement between the EWB measurements and the values recorded earlier.
18. If you are using the CD-ROM version of this book, click on the Assignments menu and select Chap. 20.

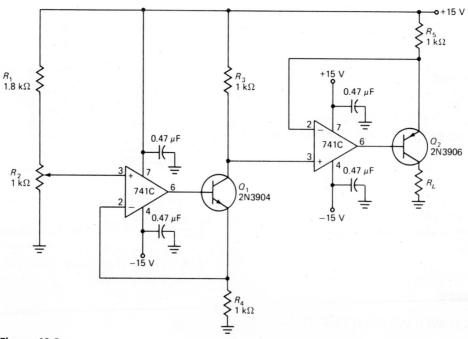

Figure 49-3

220

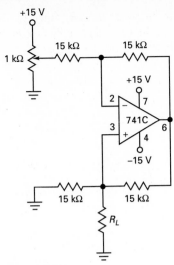

+15 V

1 kΩ 15 kΩ 15 kΩ

+15 V

2 −
7

741C
3 +
6

4

−15 V

15 kΩ 15 kΩ

R_L

Figure 49-4

ADDITIONAL WORK (OPTIONAL)

19. Connect the Howland current source of Fig. 49-4.
20. Calculate the regulated load current for an input voltage of 3 V. Record your answer:

   $I_L =$ _____ .

21. Use a DMM for voltage and current measurements. Adjust the potentiometer to get an input voltage of 3 V. Measure the load current for each of the following load resistances:

   10 Ω: $I_L =$ _____ .

   100 Ω: $I_L =$ _____ .

   1 kΩ: $I_L =$ _____ .

22. Increase the input voltage to 9 V.
23. Repeat the measurements:

   10 Ω: $I_L =$ _____ .

   100 Ω: $I_L =$ _____ .

   1 kΩ: $I_L =$ _____ .

24. Use −15 V for the potentiometer instead of +15 V. Then measure the load current for the following load resistances:

   10 Ω: $I_L =$ _____ .

   100 Ω: $I_L =$ _____ .

   1 kΩ: $I_L =$ _____ .

25. Describe the action of a Howland current source. Why is it more useful than the current source of Fig. 49-3?

Data for Experiment 49

TABLE 49-1. VOLTAGE FOLLOWER (WITHOUT CURRENT BOOSTER): V_{in} = _____

R_L	1 kΩ	330 Ω	100 Ω	33 Ω	10 Ω
V_L					
I_L					

TABLE 49-2. VOLTAGE FOLLOWER (WITH CURRENT BOOSTER): V_{in} = _____

R_L	1 kΩ	330 Ω	100 Ω	33 Ω	10 Ω
V_L					
I_L					

TABLE 49-3. VOLTAGE-CONTROLLED CURRENT SOURCE:
R_L = 100 Ω

V_{in}	Calculated I_L	Measured I_L
0 V		
1 V		
2 V		
3 V		
4 V		
5 V		

TABLE 49-4. VOLTAGE-CONTROLLED CURRENT SOURCE:
R_L = 2.2 KΩ

0 V
1 V
2 V
3 V
4 V
5 V

TABLE 49-5. TROUBLESHOOTING

Trouble	Estimated I_L	Measured I_L
R_3 short		
R_4 open		
Q_1 open		
Q_2 open		

TABLE 49-6. CRITICAL THINKING

$R =$

$I_L =$

Questions for Experiment 49

1. In Table 49-1, the maximum load current is closest to: ( )
   (a) 1.5 mA;   (b) 25 mA;   (c) 50 mA;   (d) 150 mA.
2. In Table 49-2, the load current for a 10-Ω resistor is closest to: ( )
   (a) 1.5 mA;   (b) 25 mA;   (c) 50 mA;   (d) 150 mA.
3. The measured data of Tables 49-3 and 49-4 indicate that the load resistance is   ( )
   driven by a:
   (a) voltage source;   (b) current source;   (c) transistor;   (d) op amp.
4. The last measured entry of Table 49-4 indicates that the load voltage exceeds: ( )
   (a) V_{CC};   (b) $V_{CC} - v_{in}$;   (c) V_{EE};   (d) $I_{out(max)}$.
5. As R_L increases in Fig. 49-3, the maximum input voltage: ( )
   (a) decreases;   (b) increases;   (c) stays the same;   (d) equals zero.
6. Explain why the circuit of Fig. 49-3 cannot produce 5 mA when R_L is 2.2 kΩ.

TROUBLESHOOTING

7. Why does the load current decrease to zero when R_3 is shorted in Fig. 49-3?

CRITICAL THINKING

8. Somebody has built the circuit of Fig. 49-2 with a 2N3904 instead of a 2N3055. The circuit will not work with a 10-Ω load. What is the problem?

9. Optional. Instructor's question.

10. Optional. Instructor's question.

Active Low-Pass Filters

The low-pass filter is the prototype for filter analysis and design. Computer programs convert any filter design problem into an equivalent low-pass design problem. After the computer finds the low-pass components, it converts them into equivalent components for other responses. The low-pass filter is a basic introduction to filters. Once you have built and tested some low-pass filters, you can more easily understand other filters.

In this experiment, you will build first-order low-pass filters (one-pole), second-order low-pass filters (two-pole), and an optional fourth-order low-pass filter (four-pole). To keep the wiring simple, you will work with Sallen-Key filters. This class of filters produces excellent results in a laboratory when part selection is limited. The more complicated filters require too many nonstandard resistance and capacitance values.

Besides measuring the cutoff frequency and rolloff rate of Butterworth, Bessel, and Chebyshev filters, you will look at the step response of these filters. Recall that the step response is important in digital communications. The Bessel filter has the slowest rolloff but the best step response. Of the three, the Chebyshev filter has the fastest rolloff but the worst step response. The Butterworth filter is good compromise, giving good rolloff and good step response.

Because active filters may use nonstandard capacitances, it is sometimes necessary to connect two standard capacitors in parallel to get a nonstandard value. For instance, to get 4.4 nF, use two 2.2-nF capacitors in parallel. The same idea applies to any circuit that does not have the required capacitance value: Put capacitors in parallel as needed to get as close as possible to the theoretical value.

Required Reading

Chapter 21 (Secs. 21-1 to 21-6) of *Electronic Principles*, 6th ed.

Equipment

- 1 audio generator
- 2 power supplies: ± 15 V
- 1 VOM (analog or digital multimeter)
- 1 op amp: 741C
- 13 ½-W resistors: 1 kΩ, 3.9 kΩ, two 4.7 kΩ, 8.2 kΩ, two 10 kΩ, two 33 kΩ, two 39 kΩ, two 47 kΩ
- 8 capacitors: three 2.2 nF, 3 nF, two 4.7 nF, 10 nF, 68 nF
- 1 oscilloscope

Procedure

BUTTERWORTH FIRST-ORDER LP FILTER

1. In Fig. 50-1, if you calculate the cutoff frequency and the attenuation one decade above f_C, you will get the following values:

$$f_C = 1026 \text{ Hz}$$
$$A_{db} = 20 \text{ dB}$$

2. Connect the circuit of Fig. 50-1.
3. Use channel 1 of the oscilloscope to look at the input signal (left end of 33 kΩ), and channel 2 for the output signal. Adjust the input signal to 1 V pp and 100 Hz.
4. Since the filter has unity gain in the passband, the output voltage should be equal to the input voltage.

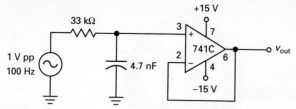

Figure 50-1

5. Find the cutoff frequency by increasing the generator frequency until the output signal is down 3 dB from the input signal. (This means that the output signal equals 0.707 V pp when the input equals 1 V pp.)

6. Increase the generator frequency to 10 times the cutoff frequency. Measure the output voltage. Record the cutoff frequency and the attenuation one decade above the cutoff frequency:

$f_C =$ _____ .

$A_{db} =$ _____ .

7. The data in the preceding step should be reasonably close to the theoretical values of Step 1.

8. Use a square-wave input. Adjust the input signal to get 1 V pp and 100 Hz. Use a time base of 1 ms/DIV. Channel 1 will display the input square wave, and channel 2 will show the step response.

9. Since this is a Butterworth first-order filter, there is no overshoot or ringing on either the positive or the negative step. Section 16-9 in the textbook discusses the risetime-bandwidth relationship of any first-order response.

BUTTERWORTH FIRST-ORDER LP FILTER WITH GAIN

10. In Fig. 50-2, calculate the cutoff frequency and attenuation one decade above cutoff:

$f_C =$ _____ .

$A_{db} =$ _____ .

11. Because the voltage gain is ideally 5.7, an input signal of approximately 175 mV pp will produce an output

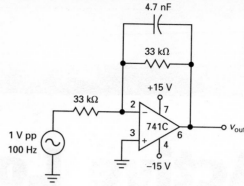

Figure 50-3

signal of 1 V pp. Connect the circuit and adjust the audio generator as needed to get an output of 1 V pp and 100 Hz.

12. Measure the cutoff frequency and attenuation one decade above the cutoff frequency. The values you record should agree with those calculated in Step 10.

$f_C =$ _____ .

$A_{db} =$ _____ .

BUTTERWORTH FIRST-ORDER INVERTING LP FILTER

13. Theoretically, the filter of Fig. 50-3 has the same cutoff frequency and rolloff as the two preceding filters. Connect the circuit and verify the cutoff frequency and attenuation one decade above cutoff:

$f_C =$ _____ .

$A_{db} =$ _____ .

BUTTERWORTH SECOND-ORDER LP FILTER

14. In Fig. 50-4, calculate the cutoff frequency, attenuation one decade above cutoff, and the Q.

$f_C =$ _____ .

$A_{db} =$ _____ .

$Q =$ _____ .

15. Based on the data in Step 14, state why the filter has a Butterworth response:

16. Connect the circuit using two 2.2 nF in parallel for the 4.4 nF. Find the cutoff frequency and measure the attenuation one decade above cutoff:

$f_C =$ _____ .

$A_{db} =$ _____ .

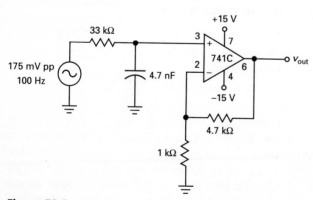

Figure 50-2

226

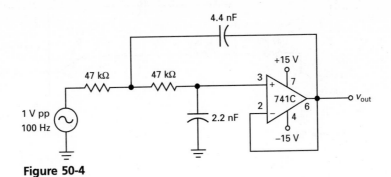

Figure 50-4

17. Use a square-wave input as you did in Step 8 to look at the output step response. This time, you will see some overshoot, as discussed in the textbook.

CHEBYSHEV SECOND-ORDER LP FILTER

18. In Fig. 50-5, calculate the cutoff frequency, attenuation one decade above cutoff, and the Q.

$f_C =$ _____ .

$A_{db} =$ _____ .

$Q =$ _____ .

19. Based on the data in Step 18 and Fig. 21-25 in the textbook, state why the filter has a Chebyshev response:

20. Connect the circuit. (*Note:* If you do not have a 68-nF capacitor, use 47 nF in parallel with 22 nF.) Find the cutoff frequency and measure the attenuation one decade above cutoff:

$f_C =$ _____ .

$A_{db} =$ _____ .

21. Use a square-wave input, as you did earlier, to look at the output step response. This time, you will see more

overshoot and ringing than you saw with the second-order Butterworth filter.

22. Sketch the output waveform superimposed on the input waveform to show the overshoot and ringing of the Chebyshev filter:

BESSEL SECOND-ORDER LP FILTER

23. In Fig. 50-6, calculate the cutoff frequency, attenuation one decade above cutoff, and the Q.

$f_C =$ _____ .

$A_{db} =$ _____ .

$Q =$ _____ .

24. Based on the data in Step 23, state why the filter has a Bessel response:

25. Connect the circuit. Find the cutoff frequency and measure the attenuation one decade above cutoff:

$f_C =$ _____ .

$A_{db} =$ _____ .

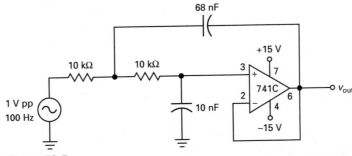

Figure 50-5

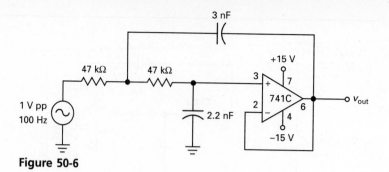

Figure 50-6

26. Use a square-wave input, as you did earlier, to look at the output step response. This time, you will see no overshoot because the Bessel has the best step response of any filter. (If there is any overshoot, the tolerance of the parts may account for it.)

27. Sketch the output waveform superimposed on the input waveform to show the step response of the Bessel filter:

30. Connect the circuit. Find the cutoff frequency and measure the attenuation one decade above cutoff:

   $f_C = $ _____ .

   $A_{db} = $ _____ .

31. Write an essay summarizing what you have learned. Start with what you consider the most important idea. Then, go on to a second idea, a third idea, and so on, until you believe that you have reinforced all the basic ideas you want to retain about active low-pass filters:

BUTTERWORTH EQUAL-COMPONENT SECOND-ORDER LP FILTER

28. In Fig. 50-7, calculate the cutoff frequency, attenuation one decade above cutoff, and the Q.

   $f_C = $ _____ .

   $A_{db} = $ _____ .

   $Q = $ _____ .

29. Based on the data in Step 28, state why the filter has a Butterworth response:

COMPUTER (OPTIONAL)

32. Repeat Steps 1 to 30 using EWB or an equivalent circuit simulator. Do not record any new values. But make sure that you get reasonable agreement between the EWB measurements and the values recorded earlier.

33. If you are using the CD-ROM version of this book, click on the Assignments menu and select Chap. 21.

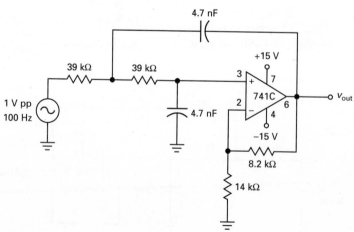

Figure 50-7

ADDITIONAL WORK (OPTIONAL)—
BUTTERWORTH FOURTH-ORDER LP FILTER

34. When the output of the upper stage of Fig. 50-8 is connected to point *A*, the overall filter has a fourth-order Butterworth response. The two stages ideally have the same cutoff frequency but the *Q*s are staggered to maintain a maximally flat response.

35. Calculate the cutoff frequency and *Q* of each stage:

$f_{C1} =$ _____ .

$Q_1 =$ _____ .

$f_{C2} =$ _____ .

$Q_2 =$ _____ .

36. Connect the circuit. Find the cutoff frequency and measure the attenuation one decade above cutoff:

$f_C =$ _____ .

$A_{db} =$ _____ .

37. Use a square-wave input as you did earlier to look at the output step response. Sketch the step response:

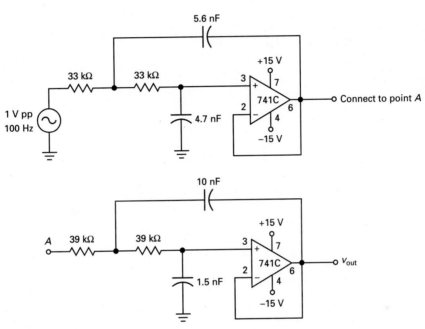

Figure 50-8

Active Butterworth Filters

The preceding experiment focused on Butterworth, Chebyshev, and Bessel responses. In this experiment, the focus is on the Butterworth response of low-pass, high-pass, wide bandpass, and narrow bandpass filters. The low-pass and high-pass filters will be Sallen-Key designs with a two-pole Butterworth response. The wide bandpass filter will be a cascade of a one-pole low-pass filter and a one-pole high-pass filter. The narrow bandpass filter will be a multiple-feedback filter with a two-pole Butterworth response.

Incidentally, the use of nanofarads has become widespread in active-filter design. Although many catalogs of electronic parts may list capacitance values like 0.001, 0.01, and 0.1 μF, most active-filter designers prefer using 1, 10, and 100 nF. Because of this, schematic diagrams of active filters typically use nanofarads when the capacitance is between 1 and 999 nF (equivalent to 0.001 and 0.999 μF).

Required Reading

Chapter 21 (Secs. 21-1 to 21-10) of *Electronic Principles*, 6th ed.

Equipment

1 audio generator
2 power supplies: ± 15 V
1 VOM (analog or digital multimeter)
2 op amp: 741C
9 ½-W resistors: 120 Ω, three 12 kΩ, two 22 kΩ, 33 kΩ, 39 kΩ, and 47 kΩ
6 capacitors: 1 nF, three 4.7 nF, two 33 nF
1 oscilloscope

Procedure

BUTTERWORTH SECOND-ORDER LOW-PASS FILTER

1. Calculate the cutoff frequency in Fig. 51-1. Record the value here:

$f_c =$ _____ .

2. What are the approximate attenuations one decade above and one decade below the cutoff frequency?

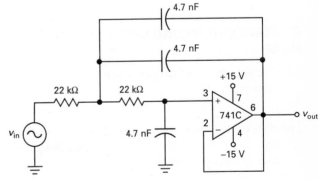

Figure 51-1

Record the values here:

Above = _____   Below = _____ .

3. Connect the circuit of Fig. 51-1.
4. Use channel 1 of the oscilloscope to look at the input signal and channel 2 for the output signal. Adjust the input signal to 1 V pp and 100 Hz.
5. Since the filter has unity gain in the passband, the output voltage should be equal to the input voltage.
6. Find the cutoff frequency by increasing the generator frequency until the output signal is down 3 dB from the input signal. (This means that the output signal

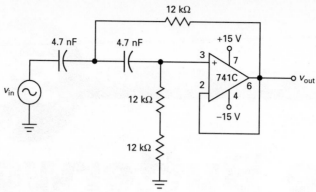

Figure 51-2

equals 0.707 V pp when the input equals 1 V pp.) Record the cutoff frequency here:

$f_c = $ _____ .

7. Increase the generator frequency to 10 times the cutoff frequency. Measure and record the peak-to-peak output voltage:

$v_{out} = $ _____ .

8. Decrease the generator frequency to one-tenth of the cutoff frequency. Measure and record the peak-to-peak output voltage:

$v_{out} = $ _____ .

9. The data in the preceding step should be reasonably close to the theoretical values of Steps 1 and 2.

10. Measure and record the output voltage for each of the input frequencies shown in Table 51-1.

BUTTERWORTH SECOND-ORDER HIGH-PASS FILTER

11. Calculate the cutoff frequency in Fig. 51-2. Record the value here:

$f_c = $ _____ .

12. What are the approximate attenuations one decade below and one decade above the cutoff frequency? Record the values here:

Below = _____ Above = _____ .

13. Connect the circuit of Fig. 51-2.

14. Use channel 1 of the oscilloscope to look at the input signal, and channel 2 for the output signal. Adjust the input signal to 1 V pp and 20 kHz.

15. Since the filter has unity gain in the passband, the output voltage should be equal to the input voltage.

16. Find the cutoff frequency by decreasing the generator frequency until the output signal is down 3 dB from the input signal. Record the cutoff frequency here:

$f_c = $ _____ .

17. Decrease the generator frequency to one-tenth of the cutoff frequency. Measure and record the peak-to-peak output voltage:

$v_{out} = $ _____ .

18. Increase the generator frequency to 10 times the cutoff frequency. Measure and record the peak-to-peak output voltage:

$v_{out} = $ _____ .

19. The data in the preceding step should be reasonably close to the theoretical values of Steps 11 and 12.

20. Measure and record the output voltage for each of the input frequencies shown in Table 51-2.

BUTTERWORTH WIDE BANDPASS FILTER

21. Calculate the lower and upper cutoff frequencies in Fig. 51-3. Record the values here:

$f_1 = $ _____ $f_2 = $ _____ .

22. Calculate the geometric center frequency and record the value here:

$f_0 = $ _____ .

23. What are the approximate attenuations one decade below the lower cutoff frequency and one decade above

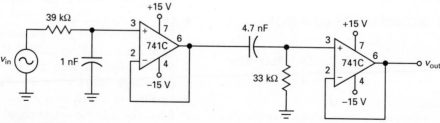

Figure 51-3

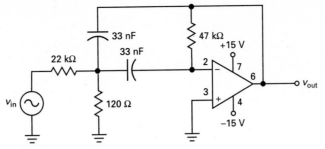

Figure 51-4

the upper cutoff frequency? Record the values here:

Below = _____ Above = _____ .

24. Connect the circuit of Fig. 51-3.
25. Use channel 1 of the oscilloscope to look at the input signal, and channel 2 for the output signal.
26. Adjust the input signal to 1 V pp and 2 kHz.
27. Find the lower cutoff frequency by decreasing the generator frequency until the output signal is down 3 dB from the input signal. Record the lower cutoff frequency here:

$f_1 =$ _____ .

28. Find the upper cutoff frequency by increasing the generator frequency until the output signal is down 3 dB from the input signal. Record the upper frequency here:

$f_2 =$ _____ .

29. The two cutoff frequencies of Steps 27 and 28 should be reasonably close to the theoretical values of Step 21.
30. Measure and record the output voltage for each of the input frequencies shown in Table 51-3.

BUTTERWORTH NARROWBAND BANDPASS FILTER

31. Calculate the center frequency and bandwidth in Fig. 51-4. Record the values here:

$f_0 =$ _____ $BW =$ _____ .

32. Connect the circuit of Fig. 51-4.
33. Use channel 1 of the oscilloscope to look at the input signal, and channel 2 for the output signal.

34. Adjust the input signal to 1 V pp and 2 kHz.
35. Find the center frequency by varying the generator frequency until the output signal is maximum. Record the center frequency here:

$f_0 =$ _____ .

36. Find the lower cutoff frequency by decreasing the generator frequency until the output signal is down 3 dB. Record the lower frequency here:

$f_1 =$ _____ .

37. Find the upper cutoff frequency by increasing the generator frequency until the output signal is down 3 dB. Record the upper frequency here:

$f_2 =$ _____ .

38. What is the bandwidth? Record the value here:

$BW =$ _____ .

39. Measure and record the output voltage for each of the input frequencies shown in Table 51-4.

COMPUTER (OPTIONAL)

40. Repeat Steps 1 to 39 using EWB or an equivalent circuit simulator. Do not record any new values. But make sure that you get reasonable agreement between the EWB measurements and the values recorded earlier.
41. If you are using the CD-ROM version of this book, click on the Assignments menu and select Chap. 21.

Data for Experiment 51

TABLE 51-1. TWO-POLE LOW-PASS RESPONSE

Frequency	Output Voltage
100 Hz	
200 Hz	
400 Hz	
1 kHz	
2 kHz	
4 kHz	
10 kHz	
20 kHz	
40 kHz	

TABLE 51-2. TWO-POLE HIGH-PASS RESPONSE

Frequency	Output Voltage
100 Hz	
200 Hz	
400 Hz	
1 kHz	
2 kHz	
4 kHz	
10 kHz	
20 kHz	
40 kHz	

TABLE 51-3. WIDE BANDPASS RESPONSE

Frequency	Output Voltage
100 Hz	
200 Hz	
400 Hz	
1 kHz	
2 kHz	
4 kHz	
10 kHz	
20 kHz	
40 kHz	

TABLE 51-4. NARROW BANDPASS RESPONSE

Frequency	Output Voltage
100 Hz	
200 Hz	
400 Hz	
1 kHz	
2 kHz	
4 kHz	
10 kHz	
20 kHz	
40 kHz	

Questions for Experiment 51

1. The circuit of Fig. 51-1 has a cutoff frequency closest to: ( )
   (a) 100 Hz;   (b) 1 kHz;   (c) 2 kHz;   (d) 10 kHz.
2. The circuit of Fig. 51-1 has a Butterworth response because it: ( )
   (a) is a low-pass filter;   (b) has equal resistors;   (c) has a Q of 0.707;
   (d) has a cutoff frequency.
3. The circuit of Fig. 51-2 has what kind of response?   ( )
   (a) low-pass;   (b) high-pass;   (c) wide bandpass;   (d) narrow bandpass.
4. The circuit of Fig. 51-2 has a cutoff frequency closest to: ( )
   (a) 100 Hz;   (b) 1 kHz;   (c) 2 kHz;   (d) 10 kHz.
5. The wide bandpass filter of Fig. 51-3 has a lower cutoff frequency closest to: ( )
   (a) 100 Hz;   (b) 1 kHz;   (c) 2 kHz;   (d) 4 kHz.
6. The wide bandpass filter of Fig. 51-3 has an upper cutoff frequency closest to: ( )
   (a) 100 Hz;   (b) 1 kHz;   (c) 2 kHz;   (d) 4 kHz.
7. The narrow bandpass filter of Fig. 51-4 has a center frequency closest to: ( )
   (a) 100 Hz;   (b) 1 kHz;   (c) 2 kHz;   (d) 4 kHz.
8. The narrow bandpass filter of Fig. 51-4 has a Q closest to: ( )
   (a) 1;   (b) 2;   (c) 4;   (d) 10.
9. Optional. Instructor's question.

10. Optional. Instructor's question.

52

Active Diode Circuits and Comparators

With op amps we can reduce the effect of diode knee voltage. The effective knee voltage is reduced by the open-loop gain of the op amp. For a typical 741C, this means the equivalent knee voltage is only 7 μV. This allows us to build circuits that will rectify, peak-detect, limit, and clamp low-level signals.

A comparator is a circuit that can indicate when the input voltage exceeds a specific limit. With a zero-crossing detector, the trip point is zero. With a limit detector, the trip point is either a positive or negative voltage.

In this experiment you will build a variety of active diode circuits as well as a zero-crossing detector and a limit detector.

Required Reading

Chapter 22 (Secs. 21-1 to 21-9) of *Electronic Principles,* 6th ed.

Equipment

1 audio generator
2 power supplies: ± 15 V
6 $\frac{1}{2}$-W resistors: 100 Ω, 1 kΩ, 2.2 kΩ, two 10 kΩ, 100 kΩ
1 potentiometer: 1 kΩ
1 diode: 1N4148 or 1N914
2 LEDs: L53RD and L53GD (or similar red and green LEDs)
1 op amp: 741C
3 capacitors: two 0.47 μF, 100 μF (rated at least 15 V)
1 VOM (analog or digital multimeter)
1 oscilloscope

Procedure

HALF-WAVE RECTIFIER

1. Build the circuit of Fig. 52-1.
2. Connect the oscilloscope (dc input) across the 10-kΩ load resistor. Set the generator to 100 Hz and adjust the level to get a peak output of 1 V on the oscilloscope. (You should be looking at a half-wave signal.)

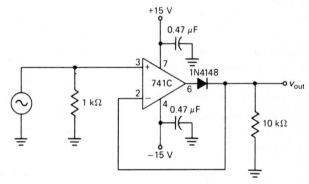

Figure 52-1

3. Measure the peak value of the input sine wave. Record the input and output peak voltages in Table 52-1.
4. Adjust the signal level to get a half-wave output with a peak value of 100 mV. Then measure the peak input voltage. Record the input and output peak voltages in Table 52-1.
5. Reverse the polarity of the diode. The output voltage should be a negative half-wave signal.

PEAK DETECTOR

6. Connect a 100-μF capacitor across the load to get the circuit of Fig. 52-2.

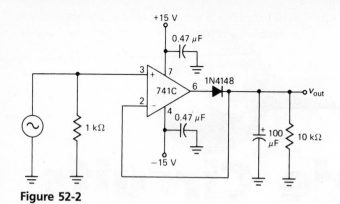

Figure 52-2

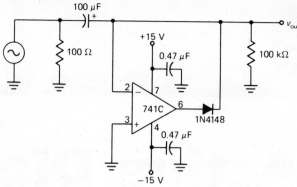

Figure 52-4

7. Adjust the generator to get an input peak value of 1 V. Measure the dc output voltage. Record the peak input voltage and the dc output voltage in Table 52-2.

8. Readjust the generator to get an input peak value of 100 mV. Measure the dc output voltage. Record the peak input voltage and the dc output voltage in Table 52-2.

9. Reverse the polarity of the diode and the capacitor. You should get a negative dc output voltage.

LIMITER

10. Build the circuit of Fig. 52-3.

11. Adjust the generator to produce a peak-to-peak value of 1 V at the left end of the 2.2-kΩ resistor.

12. Look at the output signal while turning the potentiometer through its entire range.

13. Adjust the generator to produce a peak-to-peak output of 100 mV at the left end of the 2.2-kΩ resistor. Then repeat Step 12.

14. Reverse the polarity of the diode and repeat Step 12 for a peak-to-peak output of 1 V.

DC CLAMPER

15. Connect the circuit of Fig. 52-4.

16. Adjust the input to 1 V pp.

17. Look at the output. It should be a positively clamped signal.

18. Reduce the input to 100 mV pp and repeat Step 17.

19. Reverse the polarity of the diode. The output should now be negatively clamped.

ZERO-CROSSING DETECTOR

20. Connect the zero-crossing detector of Fig. 52-5. (Note: The I_{max} out of the 741C is approximately 25 mA, so the LED current is limited to 25 mA. If an op amp has an I_{max} greater than 50 mA, you would need current-limiting resistors because most LEDs cannot handle more than 50 mA.)

21. Vary the potentiometer and notice what the LEDs do.

22. Use the dc-coupled input of the oscilloscope to look at the input voltage to pin 3. Adjust the potentiometer to get +100 mV at the input. Record the input voltage and the color of the on LED (Table 52-3).

23. Adjust the potentiometer to get an input of -100 mV. Record the input voltage and the color of the on LED.

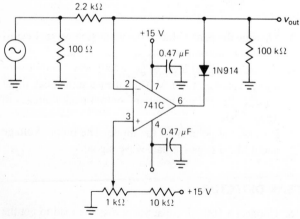

Figure 52-3

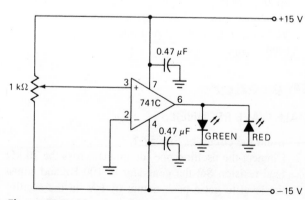

Figure 52-5

238

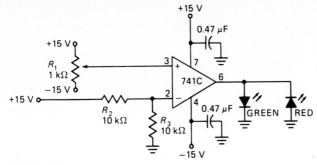

Figure 52-6

LIMIT DETECTOR

24. In Fig. 52-6, calculate the trip point of the limit detector. Record your answer in Table 52-4.
25. Connect the circuit. Adjust the input voltage until you locate the approximate trip point. Record the trip point.

TROUBLESHOOTING

26. For each set of symptoms listed in Table 52-5, try to figure out what trouble could produce the symptoms in Fig. 52-6. Insert the trouble and verify that it produces the symptoms. Record each trouble in Table 52-5.

CRITICAL THINKING

27. Select a value of R_3 to get a trip point of approximately +5 V. Connect the circuit with your design value and measure the trip point. Record the quantities listed in Table 52-6.

APPLICATION (OPTIONAL)

28. Figure 52-7 shows a window comparator similar to Fig. 22-22 in the textbook. A circuit like this can be used for go/no-go testing. If the voltage being tested is outside its normal range, we can light an LED or sound a buzzer.
29. Connect the circuit of Fig. 52-7.
30. Vary the potentiometer and notice that the red LED goes on and off.
31. With the LED on, vary the potentiometer slowly until the LED goes off. Record the input voltage:

_____ .

32. With the LED off, vary the potentiometer slowly until it goes on. Record the input voltage:

_____ .

33. Replace the LED by a buzzer (RS273-026).
34. Repeat Steps 31 and 32 for the buzzer:

$V_1 =$ _____ .

$V_2 =$ _____ .

35. What have you learned about a window comparator?

36. How can you modify the window comparator of Fig. 52-7 to test a circuit whose dc output should be between 2.5 and 3.5 V?

COMPUTER (OPTIONAL)

37. Repeat Steps 1 to 27 using EWB or an equivalent circuit simulator. Do not record any new values. But make sure that you get reasonable agreement between the EWB measurements and the values recorded earlier.
38. If you are using the CD-ROM version of this book, click on the Assignments menu and select Chap. 22.

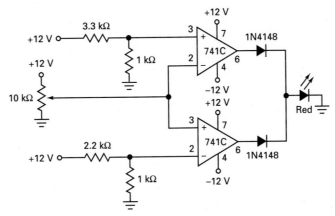

Figure 52-7

Data for Experiment 52

TABLE 52-1. ACTIVE HALF-WAVE RECTIFIER

Step 3:

$v_{in} = $ _____

$v_{out} = $ _____

Step 4:

$v_{in} = $ _____

$v_{out} = $ _____

TABLE 52-2. ACTIVE PEAK DETECTOR

Step 7:

$v_{in} = $ _____

$v_{out} = $ _____

Step 8:

$v_{in} = $ _____

$v_{out} = $ _____

TABLE 52-3. ZERO-CROSSING DETECTOR

Step 22:

$v_{in} = $ _____

Color = _____

Step 23:

$v_{in} = $ _____

Color = _____

TABLE 52-4. LIMIT DETECTOR

Calculated trip point = _____

Measured trip point = _____

TABLE 52-5. TROUBLESHOOTING

Symptoms	Trouble
1. Red LED always on	
2. Trip point equals zero	
3. Red LED goes on and off, but green LED is always off	
4. Neither LED comes on	

241

TABLE 52-6. CRITICAL THINKING

$R_3 =$ _____

Trip point = _____

Questions for Experiment 52

1. The circuit of Fig. 52-1 is a: ( )
   (a) half-wave rectifier; (b) full-wave rectifier; (c) bridge rectifier;
   (d) none of the foregoing.

2. The dc output voltage of Fig. 52-2 is approximately equal to the: ( )
   (a) peak input voltage; (b) positive supply voltage; (c) rms input
   voltage; (d) average input voltage.

3. The positive limiter of Fig. 52-3 can be adjusted to have a limiting level between ( )
   0 and approximately:
   (a) 0; (b) +1.36 V; (c) −5 V; (d) +12 V.

4. The circuit of Fig. 52-4 clamps the signal: ( )
   (a) negatively; (b) positively; (c) at −5 V; (d) at +3 V.

5. The limit detector of Fig. 52-6 has a trip point of approximately: ( )
   (a) 0; (b) +5 V; (c) +7.5 V; (d) +10 V.

6. Explain how the limit detector of Fig. 52-6 works.

TROUBLESHOOTING

7. Name at least two troubles in Fig. 52-6 that would produce a trip point of zero.

8. If the capacitor of Fig. 52-2 opens, what symptoms will you get?

CRITICAL THINKING

9. How did you arrive at your design value for R_3?

10. Optional. Instructor's question.

Waveshaping Circuits

By using positive feedback with a comparator, we can build a Schmitt trigger. It has hysteresis, which makes it less sensitive to noise. A Schmitt trigger is useful for waveshaping because it produces a square-wave output no matter what the shape of the input signal.

If we add an *RC* circuit to a Schmitt trigger, we get a relaxation oscillator. This type of circuit generates a square-wave output without an external input signal. By cascading a relaxation oscillator and an integrator, we can build a circuit that generates square waves and triangular waves.

Required Reading

Chapter 22 (Secs. 22-5 to 22-8) of *Electronic Principles,* 6th ed.

Equipment

1 audio generator
2 power supplies: adjustable from 0 to 15 V
8 ½-W resistors: 100 Ω, 1 kΩ, two 2.2 kΩ, 10 kΩ, 18 kΩ, 22 kΩ, 100 kΩ
3 op amps: 318C, two 741C
6 capacitors: two 0.1 µF, four 0.47 µF
1 oscilloscope
1 frequency counter

Procedure

SCHMITT TRIGGER

1. In Fig. 53-1, what shape do you think the output signal will have? Estimate the peak-to-peak output voltage. Record your answers in Table 53-1. Also calculate and record the trip points.
2. Connect the circuit. Adjust the input voltage to 1 V pp at 1 kHz.
3. Look at the output with an oscilloscope. Record the approximate shape of the signal in Table 53-1. Also measure and record the peak-to-peak output voltage.
4. Look at the noninverting input voltage with the oscilloscope on dc input. Measure the positive peak and record this as the UTP. Measure the negative peak and record as the LTP.

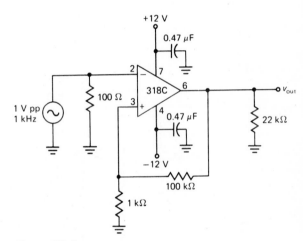

Figure 53-1

EFFECT OF SLEW-RATE LIMITING

5. Increase the frequency to 20 kHz. The output should be approximately rectangular. (*Note:* You may see some overshoot or ringing because of the high slew rate of a 318C, but the transitions between high and low should still appear almost vertical.)
6. Change the frequency back to 1 kHz. Replace the 318C by a 741C. The output should appear approximately rectangular.
7. Increase the frequency and notice how the slew rate of a 741C affects the vertical transitions.

RELAXATION OSCILLATOR AND INTEGRATOR

8. In Fig. 53-2, a relaxation oscillator drives an integrator. Calculate the frequency out of the relaxation oscillator. Record in Table 53-2.

9. Assume $+V_{sat}$ is $+14$ V and $-V_{sat}$ is -14 V. Calculate and record the peak-to-peak output voltage from the relaxation oscillator.
10. Calculate and record the peak-to-peak output voltage from the integrator.
11. Connect the circuit.
12. Look at the signal out of the relaxation oscillator. Measure its frequency and peak-to-peak value. Record these data.
13. Look at the signal at the inverting input of the relaxation oscillator. It should look like Fig. 22-32b in your textbook.
14. Look at the signal out of the integrator. Measure and record its peak-to-peak value.

TROUBLESHOOTING

15. For each trouble listed in Table 53-3, calculate the output frequency and the peak-to-peak output voltage in Fig. 53-2. Record your answers.

16. Insert each trouble into the circuit. Measure and record the output frequency and peak-to-peak voltage.

CRITICAL THINKING

17. Select a value of R_1 in Fig. 53-2 to get a frequency of approximately 1 kHz.
18. Connect the circuit with your design value of R_1. Measure the frequency. Record your value of R_1 and frequency in Table 53-4.

COMPUTER (OPTIONAL)

19. Repeat Steps 1 to 18 using EWB or an equivalent circuit simulator. Do not record any new values. But make sure that you get reasonable agreement between the EWB measurements and the values recorded earlier.
20. If you are using the CD-ROM version of this book, click on the Assignments menu and select Chap. 22.

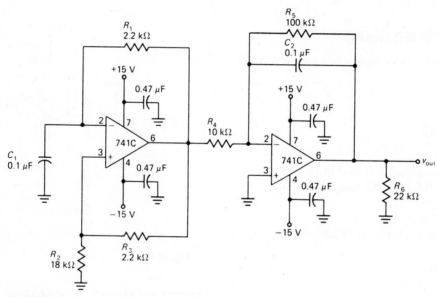

Figure 53-2

Data for Experiment 53

TABLE 53-1. SCHMITT TRIGGER

Calculated:

Shape = _____

MPP = _____

UTP = _____

LTP = _____

Measured:

Shape = _____

MPP = _____

UTP = _____

LTP = _____

TABLE 53-2. RELAXATION OSCILLATOR AND INTEGRATOR

Calculated:

f = _____

$v_{out(1)}$ = _____

$v_{out(2)}$ = _____

Measured:

f = _____

$v_{out(1)}$ = _____

$v_{out(2)}$ = _____

TABLE 53-3. TROUBLESHOOTING

| | Calculated | | Measured | |
Trouble	f	v_{out}	f	v_{out}
R_1 is 22 kΩ				
R_2 is 1.8 kΩ				
R_4 is 22 kΩ				

TABLE 53-4. CRITICAL THINKING

R_1 =

f =

Questions for Experiment 53

1. The square wave out of the Schmitt trigger (Fig. 53-1) has a peak-to-peak value   ( )
   that is closest to:
   (a) 5 V;   (b) 10 V;   (c) 20 V;   (d) 30 V.
2. The UTP of the Schmitt trigger was approximately: ( )
   (a) -0.1 V;   (b) $+0.1$ V;   (c) -10 V;   (d) $+10$ V.
3. The relaxation oscillator has a calculated frequency in Table 53-2 of: ( )
   (a) 345 Hz;   (b) 456 Hz;   (c) 796 Hz;   (d) 1.27 kHz.
4. The triangular output of the integrator in Fig. 53-2 has a calculated peak-to-peak   ( )
   value of approximately:
   (a) 0.1 V;   (b) 8.79 V;   (c) 12.3 V;   (d) 15 V.
5. The waveform at the inverting input of the relaxation oscillator (Fig. 53-2) appears: ( )
   (a) square;   (b) triangular;   (c) exponential;   (d) sinusoidal.
6. Explain how a Schmitt trigger like Fig. 53-1 works.

7. Explain how a relaxation oscillator like Fig. 53-2 works.

TROUBLESHOOTING

8. What are the symptoms in Fig. 53-2 when R_1 is 22 kΩ instead of 2.2 kΩ? Why do these
   changes occur?

CRITICAL THINKING

9. Why is it better to use a 318C instead of a 741C in a Schmitt trigger?

10. Optional. Instructor's question.

The Wien-Bridge Oscillator

The Wien-bridge oscillator is the standard oscillator circuit for low to moderate frequencies in the range of 5 Hz to about 1 MHz. The oscillation frequency is equal to $1/2\pi RC$. Typically, a tungsten lamp is used to reduce the loop gain AB to unity. It is also possible to use diodes, zener diodes, and JFETs as nonlinear elements to reduce the loop gain to unity. In this experiment you will build and test a Wien-bridge oscillator.

Required Reading

Chapter 23 (Secs. 23-1 and 23-2) of *Electronic Principles*, 6th ed.

Equipment

2   power supplies: ±15 V
9   ½-W resistors: two 1 kΩ, two 2.2 kΩ, two 4.7 kΩ, 8.2 kΩ, two 10 kΩ
3   diodes: 1N4148 or 1N914
1   LED: L53RD (or similar red LED)
1   op amp: 741C
1   potentiometer: 5 kΩ

4   capacitors: two 0.01 μF, two 0.47 μF
1   oscilloscope
1   frequency counter

Procedure

OSCILLATOR

1. In Fig. 54-1, calculate and record the frequency of oscillation for each value of R in Table 54-1. Also, calculate and record the MPP values. (*Note:* The LED is used to indicate when the circuit is oscillating. The 1N4148 across the LED protects it during reverse bias because the LED has a breakdown voltage of only 3 V.)

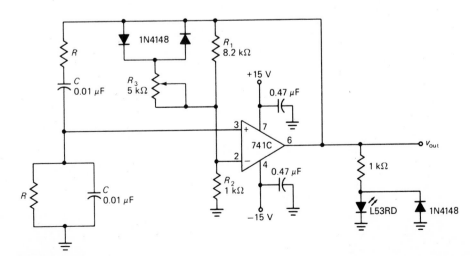

Figure 54-1

2. Connect the circuit with an R of 10 kΩ. Look at the output with an oscilloscope. Adjust R_3 to get as large an unclipped output as possible.

3. Measure the frequency. Measure the peak-to-peak output voltage. Record both quantities in Table 54-1.

4. Repeat Steps 2 and 3 for the other values of R.

TROUBLESHOOTING

5. Insert each trouble listed in Table 54-2. Determine what effect the trouble has on the output signal. Record the symptoms in Table 54-2. (Examples of entries are "no output," "heavily clipped output," and "small distorted output.")

CRITICAL THINKING

6. Select a value of R (nearest standard value) to get an oscillation frequency of approximately 2.25 kHz.

7. Connect the circuit with your value of R. Measure the frequency. Record the quantities in Table 54-3.

COMPUTER (OPTIONAL)

8. Repeat Steps 1 to 7 using EWB or an equivalent circuit simulator. Do not record any new values. But make sure that you get reasonable agreement between the EWB measurements and the values recorded earlier.

9. If you are using the CD-ROM version of this book, click on the Assignments menu and select Chap. 23.

Data for Experiment 54

TABLE 54-1. OSCILLATOR

	Calculated			Measured	
R	f		MPP	f	MPP
10 kΩ					
4.7 kΩ					
2.2 kΩ					

TABLE 54-2. TROUBLESHOOTING

Trouble	Symptoms
R_1 short	
R_1 open	
R_2 short	
R_2 open	
R_3 short	
R_3 open	

TABLE 54-3. CRITICAL THINKING

$R =$	
$f =$	

Questions for Experiment 54

1. The data of Table 54-1 indicate that an increase in resistance produces which of   ( )
   the following changes in oscillation frequency:
   (a) decrease;   **(b)** increase;   **(c)** no change.
2. The MPP value of the circuit was closest to: ( )
   (a) 0.7 V;   **(b)** 1.4 V;   **(c)** 15 V;   **(d)** 27 V.
3. The component that is involved in reducing loop gain to unity is the: ( )
   (a) 741C;   **(b)** LED;   **(c)** 1N4148;   **(d)** 0.01 μF.
4. The peak LED current is closest to: ( )
   (a) 8.59 mA;   **(b)** 17.1 mA;   **(c)** 19.1 mA;   **(d)** 27 mA.
5. The 1N4148 protects the LED from excessive reverse voltage because the 1N4148:   ( )
   (a) breaks down first;   **(b)** conducts when the reverse voltage exceeds -0.7 V;
   (c) goes into reverse bias when the LED is on;   **(d)** has a higher power dissipation than the LED.
6. Briefly explain how a Wien-bridge oscillator works.

7. Explain why there is no output when R_1 is shorted.

8. Explain why the output is heavily clipped when R_3 is open.

CRITICAL THINKING

9. How can you make the Wien-bridge oscillator tunable to different frequencies?

10. Optional. Instructor's question.

55

The LC Oscillator

For oscillation frequencies between approximately 1 and 500 MHz, the *LC* oscillator is used instead of a Wien-bridge oscillator. This type of oscillator uses a resonant *LC* tank circuit to determine the frequency. The Colpitts oscillator is the most widely used *LC* oscillator because the feedback voltage is conveniently produced by a capacitive voltage divider rather than an inductive divider (Hartley). For an oscillator to start, the small-signal voltage gain must be greater than the reciprocal of the feedback fraction. In symbols, $A > 1/B$. As the oscillations increase, the value of A decreases until the loop gain AB equals unity.

Required Reading

Chapter 23 (Sec. 23-4) of *Electronic Principles,* 6th ed.

Equipment

1 power supply: 15 V
2 ½-W resistors: 22 kΩ, 47 kΩ
1 inductor: 33 mH (or nearest value)
4 capacitors: 100 pF, 1000 pF, 0.1 μF, 0.47 μF
1 transistor
1 potentiometer: 50 kΩ
1 oscilloscope
1 frequency counter

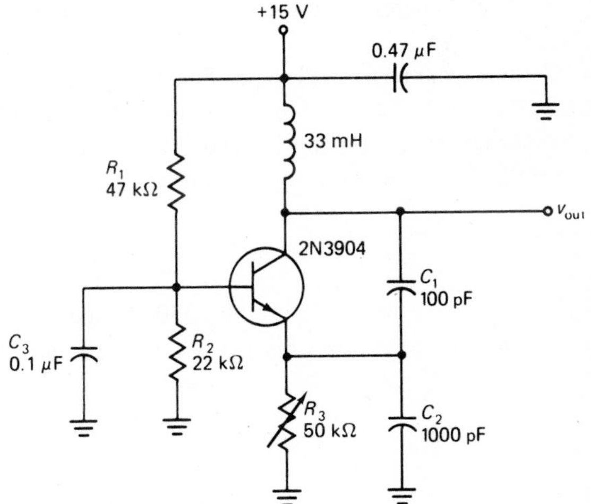

Figure 55-1

Procedure

COLPITTS OSCILLATOR

1. In Fig. 55-1, neglect transistor and stray-wiring capacitance. Calculate the frequency of oscillation. Also calculate the peak-to-peak output voltage and the feedback fraction. Record your answers in Table 55-1. (*Note:* The 0.47-μF capacitor is a supply bypass capacitor needed with some power supplies. This capacitor ac grounds the upper end of the 33 mH and prevents the supply impedance from affecting the oscillation frequency and amplitude.)
2. Connect the circuit.
3. Look at the emitter signal with channel 1 and the output signal with channel 2 of an oscilloscope.
4. Vary the 50-kΩ emitter resistor through its range

until oscillations start. If oscillations do not start, try a 100-kΩ potentiometer.
5. Adjust this resistor so that the output signal is just starting to clip.
6. Measure the frequency of oscillation and the peak-to-peak output voltage. Record the values.
7. Measure the peak-to-peak emitter signal. Calculate and record the feedback fraction B in Table 55-1.

TROUBLESHOOTING

8. Insert each trouble listed in Table 55-2. Record the output symptoms. Examples of entries are "no output," "smaller output," and "higher frequency."

253

CRITICAL THINKING

9. Ignore transistor and stray capacitance. Select new values for C_1 and C_2 to get an oscillation frequency that is 50 percent higher than the calculated frequency in Table 55-1.

10. Insert your design values. Measure the oscillation frequency. Record all quantities in Table 55-3.

COMPUTER (OPTIONAL)

11. Repeat Steps 1 to 10 using EWB or an equivalent circuit simulator. Do not record any new values. But make sure that you get reasonable agreement between the EWB measurements and the values recorded earlier.

12. If you are using the CD-ROM version of this book, click on the Assignments menu and select Chap. 23.

ADDITIONAL WORK (OPTIONAL)

13. As discussed in Experiment 27, you must be constantly aware of the loading effect of the oscilloscope probes. An inexpensive oscilloscope may add as much as 200 pF to a circuit on the X1 probe position. If you use the X10 probe position, the input capacitance decreases to 20 pF. Therefore, use the X10 position whenever possible.

14. Probe capacitances will depend on the quality of the oscilloscope you are using. Find out what the probe capacitances are for the X1 and X10 position of the oscilloscope you are using. Record the values:

$C_{(X1)} = $ _____ .

$C_{(X10)} = $ _____ .

15. Replace the 33-mH inductor by a 330-μH inductor (or an inductor near this value). Since the resonant fre-

quency is inversely proportional to the square root of inductance, the new resonant frequency will be approximately 10 times higher than before.

16. Look at the output with the X1 position of the probe. Measure the resonant frequency and record the value:

$f_r = $ _____ (X1).

17. Switch the probe to the X10 position. Increase the sensitivity by a factor of 10 to compensate for the probe attenuation of X10.

18. Repeat Step 16 and record the resonant frequency:

$f_r = $ _____ (X10).

19. The frequency in Step 18 should be slightly higher than the frequency in Step 16 because the probe will add less capacitance to the circuit during the measurement of resonant frequency.

20. When no probe is connected, is the resonant frequency higher, lower, or the same as the frequency in Step 18?

Answer = _____ .

21. What have you learned about the effect an oscilloscope probe has on a circuit under test conditions?

22. If you were to shorten all the wires as much as possible in your circuit, what effect would this have on the resonant frequency? Why do you say this?

Data for Experiment 55

TABLE 55-1. COLPITTS OSCILLATOR

Calculated	Measured
$f =$	$f =$
MPP $=$	MPP $=$
$B =$	$B =$

TABLE 55-2. TROUBLESHOOTING

Trouble	Output Symptoms
R_1 short	
R_1 open	
R_2 short	
R_2 open	
R_3 short	
R_3 open	
C_1 short	
C_1 open	
C_2 short	
C_2 open	
C_3 short	
C_3 open	

TABLE 55-3. CRITICAL THINKING

$C_1 =$

$C_2 =$

$f =$

Questions for Experiment 55

1. The calculated oscillation frequency of Fig. 55-1 is approximately:
   (a) 90 kHz; (b) 225 kHz; (c) 445 kHz; (d) 528 kHz. ( )
2. The calculated feedback fraction of Fig. 55-1 is closest to:
   (a) 0.091; (b) 0.1; (c) 1; (d) 10. ( )
3. For the oscillator to start, the minimum voltage gain is approximately:
   (a) 1; (b) 5; (c) 11; (d) 25. ( )
4. The LC oscillator of Fig. 55-1 is an example of a:
   (a) CB oscillator; (b) CE oscillator; (c) Wien-bridge oscillator; ( )
   (d) twin-T oscillator.

5. The calculated peak-to-peak output voltage of Fig. 55-1 is approximately: ( )
   (a) 20 V;   **(b)** 25 V;   **(c)** 30 V;   **(d)** 40 V.

6. Briefly explain how an *LC* oscillator works.

7. In Fig. 55-1, what effect will transistor and stray-wiring capacitance have on the frequency of oscillation?

TROUBLESHOOTING

8. Why is there no output when R_1 is open?

CRITICAL THINKING

9. Explain why the actual frequency of oscillation will be less than your calculated frequency of oscillation.

10. Optional. Instructor's question.

Op-Amp Applications: Signal Generators

A function generator is a widely used instrument that produces three output shapes: sinusoidal, rectangular, and triangular. Figure 56-1 shows a simple function generator that you will build in this experiment. The first stage is a Wien-bridge oscillator, the second stage is a Schmitt trigger, and the third stage is an integrator.

You will build and test each stage starting with the Wien-bridge oscillator. After it is working properly, you can build and test the Schmitt trigger. Finally, you can connect the integrator and the function generator will be complete. Point *A* will have sinusoidal output, point *B* will have a rectangular output, and point *C* will have a triangular output.

Required Reading

Chapters 22 and 23 (Secs. 22-3, 22-5, and 23-2) of *Electronic Principles*, 6th ed.

Equipment

- 2   power supplies: ±15 V
- 1   VOM (analog or digital multimeter)
- 4   diodes: three 1N4148, L53RD
- 4   op amps: three LM741C, LM318
- 10   ½-W resistors: two 1 kΩ, 2.2 kΩ, 8.2 kΩ, 10 kΩ, two 15 kΩ, 18 kΩ, 22 kΩ, 100 kΩ
- 5   capacitors: 0.0068 μF, two 0.01 μF, 0.022 μF, 0.047 μF
- 1   potentiometer: 5 kΩ
- 1   oscilloscope

Procedure

WIEN-BRIDGE OSCILLATOR

1. Connect the Wien-bridge oscillator.
2. Look at the output of pin 6 with channel 1 of the oscilloscope (5 V/DIV). Adjust the variable 5-kΩ resistor to get a maximum unclipped sinusoidal output.
3. The LED should be lit, indicating an output signal.
4. Measure and record the frequency and the peak-to-peak output voltage:

    $f_{\text{out}} =$ _____ .

    $v_{\text{out}} =$ _____ .

5. Sketch the output waveform of the Wien-bridge oscillator, showing the peak-to-peak value and the period:

SCHMITT TRIGGER

6. Connect the Schmitt trigger. This includes all the circuitry between points *A* and *B*.
7. Test the Schmitt trigger as follows: Connect the output of the Wien-bridge oscillator to point *A*. Use channel 1 to monitor the input to the Schmitt trigger (point *A*), and channel 2 to monitor the output of the Schmitt trigger (point *B*).

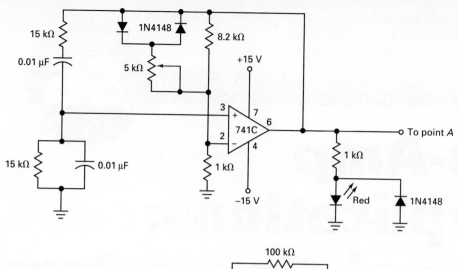

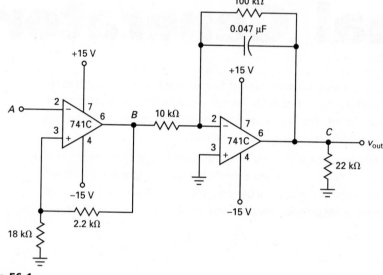

Figure 56-1

8. The display on the oscilloscope should show a sine wave and a square wave. Both will have approximately the same peak-to-peak value.

9. Record the frequency and peak-to-peak value of the square wave:

   $f_{out} =$ _____ .

   $v_{out} =$ _____ .

10. Sketch the input and output waveforms of the Schmitt trigger, showing the peak-to-peak values and the period:

INTEGRATOR

11. Connect the integrator.

12. Test the integrator as follows: Connect the output of the Schmitt trigger to the input of the integrator, if you have not already done so.

13. Use channel 1 to monitor the input to the integrator (point *B*) and channel 2 to monitor the output of

the integrator (point *C*). You should see a triangular wave.

14. Record the frequency and peak-to-peak value of the triangular wave:

   $f_{out} =$ _____ .

   $v_{out} =$ _____ .

15. Sketch the input and output waveforms of the integrator, showing the peak-to-peak values and the period:

16. The triangular wave has a smaller peak-to-peak value than the sine wave and square wave. One way to increase its peak-to-peak value is to use a smaller capacitor.

17. Change the 0.047-μF capacitor to a 0.022-μF capacitor. Record the peak-to-peak value of the triangular wave:

   _____ .

18. Increase the frequency of the Wien-bridge oscillator by changing the two 0.01-μF capacitors to 0.0047 μF.

258

19. Notice how the slew rate of a 741C is degrading the shape of the square wave and the triangular wave. Since a 741C has a typical slew rate of 0.5 V/μs, it takes typically 56 μs to slew through a peak-to-peak change of 28 V. Since the period is theoretically 443 μs, the slew-rate distortion is noticeable.

20. Here is one way to improve the performance: Turn the power off and replace the 741C by an LM318.

21. Turn the power back on, and you will see well-defined square waves and triangular waves on the oscilloscope. Explain why the waveforms look cleaner and sharper:

CRITICAL THINKING

22. Briefly describe how you would modify the function generator of Fig. 56-1 to make it tunable over the audio range of 20 Hz to 20 kHz.

23. Suppose we want to buffer the A, B, and C outputs and have a 600-Ω output impedance on each buffered output. Describe one simple way to do it by using three more op amps:

COMPUTER (OPTIONAL)

24. Repeat Steps 1 to 21 using EWB or an equivalent circuit simulator. Do not record any new values. But make sure that you get reasonable agreement between the EWB measurements and the values recorded earlier.

25. If you are using the CD-ROM version of this book, click on the Assignments menu and select Chap. 23.

Questions for Experiment 56

1. If the supply voltages in Fig. 56-1 are changed to ± 10 V, the peak-to-peak output ( )
of the Wien-bridge oscillator is closest to:
   (a) 10 V;   (b) 20 V;   (c) 30 V;   (d) 40 V.

2. If all diodes are reversed in Fig. 56-1, the: ( )
   (a) oscillator will stop;   (b) LED will go out;   (c) LED is destroyed;
   (d) circuit works as before.

3. If the 10-kΩ resistance is reduced slightly in Fig. 56-1, the: ( )
   (a) square-wave output becomes smaller;   (b) triangular wave becomes
   bigger;   (c) LED goes out;   (d) square wave becomes smaller.

4. The Schmitt trigger works better with an LM318 because this op amp has: ( )
   (a) less offset voltage;   (b) a higher CMRR;   (c) a higher slew
   rate;   (d) an open collector.

5. Reducing the capacitors in the Wien-bridge oscillator will: ( )
   (a) decrease the frequency;   (b) increase the frequency;   (c) produce a
   faster slew rate;   (d) increase the triangular output.

6. The average current through the LED in Fig. 56-1 is closest to: ( )
   (a) 3 mA;   (b) 6 mA;   (c) 12 mA;   (d) 24 mA.

7. If the 2.2-kΩ resistor of Fig. 56-1 is increased, the: ( )
   (a) hysteresis is reduced;   (b) peak-to-peak value is increased;
   (c) frequency is decreased;   (d) output offset is reduced.

8. To operate the function generator of Fig. 56-1 at higher frequencies, we need to: ( )
   (a) shorten the wires;   (b) use a better op amp;   (c) use smaller
   capacitors;   (d) all of the foregoing.

9. Optional. Instructor's question.

10. Optional. Instructor's question.

The 555 Timer

The 555 timer combines a relaxation oscillator, two comparators, and an *RS* flip-flop. This versatile chip can be used as an astable multivibrator, monostable multivibrator, VCO, ramp generator, etc. In this experiment, you will build and test some basic 555 timer circuits.

Required Reading

Chapter 23 (Secs. 23-7 and 23-8) of *Electronic Principles,* 6th ed.

Equipment

1   audio generator
1   power supply: 15 V
10 ½-W resistors: two 1 kΩ, 4.7 kΩ, two 10 kΩ, 22 kΩ, 33 kΩ, 47 kΩ, 68 kΩ, 100 kΩ
1   potentiometer: 1 kΩ
4   capacitors: 0.01 μF, 0.1 μF, two 0.47 μF
1   transistor: 2N3906
1   op amp: 741C
1   timer: NE555
1   oscilloscope
1   frequency counter

Procedure

ASTABLE 555 TIMER

1. Calculate the frequency and duty cycle in Fig. 57-1 for the resistances listed in Table 57-1. Record your answers.
2. Connect the circuit of Fig. 57-1 with $R_1 = 10$ kΩ and $R_2 = 100$ kΩ.
3. Look at the output with an oscilloscope. Measure and record the frequency.
4. Measure *W*. Calculate and record the duty cycle as the measured *D* in Table 57-1.
5. Look at the voltage across the timing capacitor (pin 6). You should see an exponentially rising and falling wave between 5 and 10 V.
6. Repeat Steps 2 through 5 for the other resistances of Table 57-1.

VOLTAGE-CONTROLLED OSCILLATOR

7. Connect the VCO of Fig. 57-2.

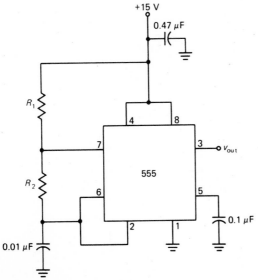

Figure 57-1

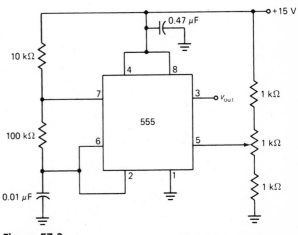

Figure 57-2

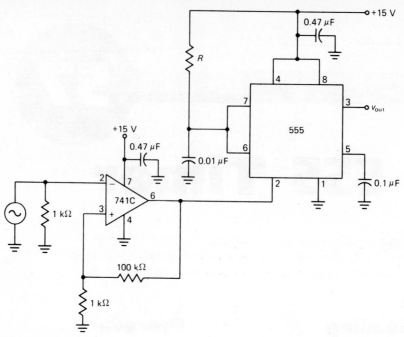

Figure 57-3

8. Look at the output with an oscilloscope. Vary the potentiometer and notice what happens. Record the minimum and maximum frequencies in Table 57-2.

9. Disconnect pin 5 from the potentiometer and couple a 10-Hz signal into this pin. Slowly increase the level of this signal from zero. Notice the horizontal jitter that appears on the signal. This is the PPM discussed in Sec. 23-9 of the textbook.

MONOSTABLE 555 TIMER

10. Figure 57-3 shows a Schmitt trigger driving a monostable 555 timer. Assume that it produces a normal

trigger input for the 555. Calculate and record the pulse width out of the 555 timer for each *R* listed in Table 57-3.

11. Connect the circuit of Fig. 57-3 with an *R* of 33 kΩ.

12. Look at the output of the Schmitt trigger (pin 6 of the 741C). Set the frequency of the sine-wave input to 1 kHz. Adjust the sine-wave level until you get a Schmitt-trigger output with a duty cycle of approximately 90 percent.

13. Look at the output of the 555 timer. Measure and record the pulse width.

14. Repeat Steps 11 through 13 for the remaining *R* values of Table 57-3.

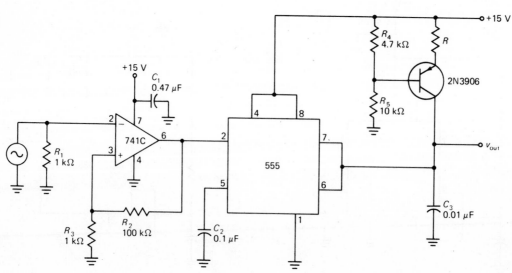

Figure 57-4

RAMP GENERATOR

15. Figure 57-4 shows a ramp generator. As before, the Schmitt trigger drives a 555 timer connected for monostable operation. But now the timing capacitor is charged by a *pnp* current source rather than a resistor. For each value of R listed in Table 57-4, calculate the slope of the output waveform.

16. Connect the circuit of Fig. 57-4 with an R of 10 kΩ.

17. Set the ac generator to 1 kHz. Adjust the level to get a duty cycle of approximately 90 percent out of the Schmitt trigger.

18. Look at the output voltage; it should be a positive ramp. Measure the ramp voltage and time. Then work out the slope. Record the value in Table 57-4.

19. Repeat Steps 16 to 18 for the remaining values of R (Table 57-4).

TROUBLESHOOTING

20. Assume that R equals 22 kΩ in Fig. 57-4. Here are the symptoms: (1) no ramp appears at the final output; (2) a normal Schmitt-trigger output drives pin 2 of the 555 timer; (3) approximately +10 V appears at the base of the 2N3904. Try to figure out what troubles (there is more than one possibility) can cause these symptoms. Insert each suspected trouble to verify that it does cause the symptoms. Record all the troubles you locate (Table 57-5).

CRITICAL THINKING

21. Select a value of C_3 that produces a slope of 15 V/ms when R equals 10 kΩ.

22. Connect the circuit with an R of 10 kΩ and your design value of C_3. Measure the slope of the output signal. Record the quantities of Table 57-6.

COMPUTER (OPTIONAL)

23. Repeat Steps 1 to 22 using EWB or an equivalent circuit simulator. Do not record any new values. But make sure that you get reasonable agreement between the EWB measurements and the values recorded earlier.

24. If you are using the CD-ROM version of this book, click on the Assignments menu and select Chap. 23.

265

Data for Experiment 57

TABLE 57-1. ASTABLE MULTIVIBRATOR

		Calculated		Measured	
R_1	R_2	f	D	f	D
10 kΩ	100 kΩ				
100 kΩ	10 kΩ				
10 kΩ	10 kΩ				

TABLE 57-2. VCO OPERATION

$f_{min} =$ _____

$f_{max} =$ _____

TABLE 57-3. MONOSTABLE MULTIVIBRATOR

R	Calculated W	Measured W
33 kΩ		
47 kΩ		
68 kΩ		

TABLE 57-4. RAMP GENERATOR

R	Calculated Slope	Measured Slope
10 kΩ		
22 kΩ		
33 kΩ		

TABLE 57-5. TROUBLESHOOTING

Trouble	Description
1	
2	
3	

TABLE 57-6. CRITICAL THINKING

$C_3 =$ _____

Slope $=$ _____

Questions for Experiment 57

1. In Fig. 57-1, the calculated frequency for R_1 and R_2 (both equal to 10 kΩ) is ( )
   approximately:
   (a) 686 Hz;   (b) 1.2 kHz;   (c) 4.8 kHz;   (d) 6.91 kHz.

2. In Fig. 57-2, the adjustment controls the: ( )
   (a) output frequency;   (b) output voltage;   (c) supply voltage;
   (d) input voltage.

3. The output of the Schmitt trigger in Fig. 57-3 is: ( )
   (a) always positive;   (b) always negative;   (c) positive on one half cycle
   and negative on the other;   (d) a constant dc voltage.

4. An R of 47 kΩ in Fig. 57-3 produces a pulse that is closest to: ( )
   (a) 363 μs;   (b) 517 μs;   (c) 748 μs;   (d) 1000 μs.

5. In Fig. 57-4, an R of 10 kΩ produces a slope of approximately: ( )
   (a) 12.4 V/ms;   (b) 18.6 V/ms;   (c) 41 V/ms;   (d) 56 V/ms.

6. In Fig. 57-3, how much input voltage do we need to get an output from the Schmitt
   trigger? Why?

7. Briefly describe the circuit operation of Fig. 57-4.

TROUBLESHOOTING

8. In Fig. 57-4, assume that R equals 10 kΩ and the output slope is 410 V/ms. Name a
   trouble that can produce this slope.

CRITICAL THINKING

9. How did you figure out your design values?

10. Optional. Instructor's question.

555 Timer Applications

This experiment will provide you with additional insight into the 555, a widely used timing circuit. The first application is a START and RESET circuit. Pushing a START button triggers a monostable circuit. The circuit can be reset at any time by pushing the RESET button. The second application is an ALARM circuit. Pushing the ALARM switch turns on an astable multivibrator that drives a speaker. The third application is a pulse-position modulator that uses two 555 timers. The first timer modulates the second, the output of which drives the speaker.

Required Reading

Chapter 23 (Secs. 23-7 and 23-9) of *Electronic Principles*, 6th ed.

Equipment

1 power supply: 12 V
1 VOM (analog or digital multimeter)
2 diodes: L53RD, LR53GD
2 timers: two LM555
11 ½-W resistors: 100 Ω, 470 Ω, 1 kΩ, 1.2 kΩ, 1.8 kΩ, three 10 kΩ, 100 kΩ, two 1 MΩ
8 capacitors: two 0.01 μF, 0.1 μF, 0.33 μF, 1 μF, two 10 μF, 470 μF
1 8-Ω speaker (small)
1 oscilloscope

Procedure

START AND RESET

1. Figure 58-1 is the START-and-RESET circuit discussed in Sec. 23-9 of the textbook. The START button controls the beginning of a timed interval of high output voltage. The RESET button can be used to terminate the interval if desired.
2. Connect the circuit of Fig. 58-1. Set the 10-kΩ variable resistance to minimum. Use push-button switches if available. If not, use SPST switches or temporary jumper connections when called for.

3. Push the START switch and release it. (If you are using SPSD switches, close the switch temporarily and then open it. If you are using jumper wires, temporarily connect a wire between pin 2 and ground and then disconnect it.)
4. The LED will light up. It will stay on for an interval of time. Then it will go out.
5. Use channel 1 to look at the charging waveform on pin 6, and channel 2 for the output on pin 3.
6. Repeat Step 3 and watch the waveforms on the oscilloscope. With a stopwatch or by counting 1001, 1002, 1003, and so on, determine how long the LED stays on. Record the time in seconds:

_____ .

7. Set the 10-kΩ variable resistor to maximum. Repeat Step 6 and record the time:

_____ .

8. Steps 8 to 13 are optional. If you have a buzzer and a motor, do these steps. Connect a buzzer (RS273-026) between pin 3 and ground. The buzzer will be in parallel with the LED branch.
9. Repeat Step 3. The buzzer will sound, and the LED will light for a timed interval. Then, both will become inactive.
10. Disconnect the buzzer and the LED branch, so that nothing is connected to pin 3.
11. Add the emitter follower and motor of Fig. 58-2 to the circuit.
12. Repeat Step 3. The motor will run for a timed interval and then stop.

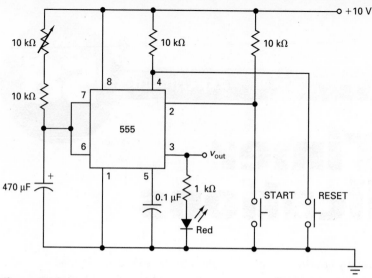

Figure 58-1

13. What did you learn in Steps 3 to 12?

20. What did you learn in Steps 14 to 19?

SIRENS AND ALARMS

14. Figure 58-3 shows a circuit that can sound an alarm when the ALARM switch is opened. It is an astable multivibrator, waiting for someone or something to open the alarm switch.
15. Connect the circuit.
16. When power is applied, open the ALARM switch, and you will hear a sound from the speaker.
17. Vary the 1-MΩ resistor to change the frequency.
18. Steps 18 to 20 are optional. If you have a photocell, do these steps. Connect a photocell (RS276-1657) in parallel with the ALARM switch. Point the photocell toward a light source.
19. Open the ALARM switch. Pass your hand over the photocell to darken it. The speaker will produce a sound. (*Note:* If the speaker does not produce a sound, the photocell resistance is too high. Try again after changing the 10-kΩ resistor on pin 4 to 100 kΩ.)

VCO WITH PULSE-POSITION MODULATION

21. In Fig. 58-4, both stages are astable mutlivibrators. Calculate the oscillation frequency for each stage:

$$f_1 = \underline{\hspace{2cm}} . \quad f_2 = \underline{\hspace{2cm}} .$$

22. With the switch in position 2, only the frequency of the second stage reaches the speaker. But when the switch is moved to position 1, the output of the first stage is used to modulate the second stage. This will produce a warbled output from the second stage, a sound that rises and falls in frequency.
23. Connect all of the circuit shown in Fig. 58-4, except the speaker.
24. After power is applied, the green LED should be flashing slowly. This indicates that stage 1 is oscillating at a very low frequency.
25. Make sure that the switch in the second stage is in position 2. Look at the output of stage 2 with an oscilloscope to make sure that it is oscillating at approximately 900 Hz. After synchronizing the signal, you should see a rectangular wave with a duty cycle of about 60 percent.
26. Continue to monitor the output of the second stage with the oscilloscope. Change the switch in the second stage to position 1, so that the output of the first stage can modulate the output of the second stage.
27. State your observations about the modulated output frequency as seen on the oscilloscope:

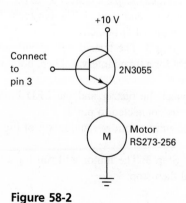

Figure 58-2

270

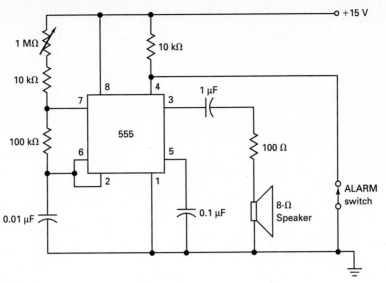

Figure 58-3

28. Move the switch back to position 2 to remove the modulation. Connect a small 8-Ω speaker to the output, as shown in Fig. 58-4. You should hear a speaker output with a frequency of approximately 900 Hz.

29. Move the switch to position 1 to get modulation. Now, you should hear a warbled output sound.

COMPUTER (OPTIONAL)

30. If you are using the CD-ROM version of this book, click on the Assignments menu and select Chap. 23.

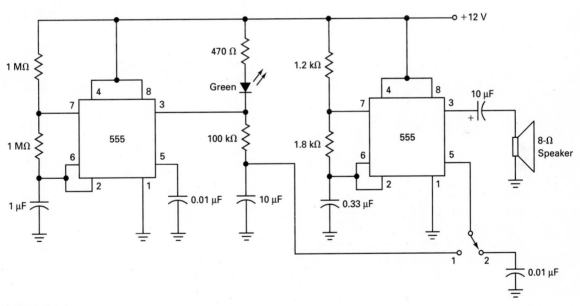

Figure 58-4

Questions for Experiment 58

1. The circuit of Fig. 58-1 is: ( )
   (a) monostable;   (b) astable;   (c) bistable;   (d) unstable.
2. The circuit of Fig. 58-1 has a maximum pulse width of approximately: ( )
   (a) 1 s;   (b) 2 s;   (c) 5 s;   (d) 10 s.
3. If the motor of Fig. 58-2 has a current of 1 A, the base current is closest to: ( )
   (a) 1 μA;   (b) 2 mA;   (c) 200 mA;   (d) 1 A.
4. The circuit of Fig. 58-3 is: ( )
   (a) monostable;   (b) astable;   (c) bistable;   (d) ultrastable.
5. The circuit of Fig. 58-3 has a maximum frequency of approximately: ( )
   (a) 123 Hz;   (b) 686 Hz;   (c) 5 kHz;   (d) 11.2 kHz.
6. The first stage of Fig. 58-4 has an output frequency of approximately: ( )
   (a) 0.46 Hz;   (b) 2.3 Hz;   (c) 123 Hz;   (d) 909 Hz.
7. The second stage of Fig. 58-4 has an output frequency of approximately: ( )
   (a) 0.48 Hz;   (b) 2.3 Hz;   (c) 454 Hz;   (d) 909 Hz.
8. When the LED is turned on in Fig. 58-4, the LED current is approximately: ( )
   (a) 10 mA;   (b) 20 mA;   (c) 40 mA;   (d) 80 mA.
9. If both timing resistances are doubled in the second stage of Fig. 58-4, the output   ( )
   frequency is closest to:
   (a) 0.48 Hz;   (b) 2.3 Hz;   (c) 454 Hz;   (d) 909 Hz.
10. Optional. Instructor's question.

59

Shunt Regulators

The shunt regulator works fine as long as the changes in line and load are not too great. It has the advantage of a simple circuit design that includes built-in short-circuit protection. It has the disadvantage of poor efficiency because of the power wasted in the series resistor. In this experiment, you will build and test three different shunt regulators.

Required Reading

Chapter 24 (Secs. 24-1 and 24-2) of *Electronic Principles*, 6th ed.

Equipment

- 2 power supplies: ±15 V
- 1 VOM (analog or digital multimeter)
- 2 diodes: 1N753, 1N757
- 2 transistors: 2N3904, 2N3055
- 1 op amp: 741C
- 14 ½-W resistors: two 100 Ω, 220 Ω, 470 Ω, two 1 kΩ, three 330 Ω, 470 Ω, 2.2 kΩ, 3.3 kΩ, 5.6 kΩ, 6.8 kΩ
- 2 potentiometers: 1 kΩ, 5 kΩ
- 1 oscilloscope

Procedure

FIXED SHUNT REGULATOR

1. In Fig. 59-1, half-watt resistors are used in parallel to avoid the need for 1-W resistors. The parallel 100-Ω resistors produce an equivalent series resistance of 50 Ω, and the parallel 330-Ω resistors produce an equivalent load resistance of 110 Ω.

2. Connect the circuit of Fig. 59-1.

3. Use a DMM to measure the output voltage across the 330-Ω resistors. Record this value as the full-load voltage:

   $V_{FL} = $ _____ .

4. Remove one of the 330-Ω resistors and record the output voltage:

   _____ .

5. Remove another 330-Ω resistor and record the output voltage:

   _____ .

6. Remove the last 330-Ω resistor and record this as the no-load voltage:

   $V_{NL} = $ _____ .

7. Calculate and record the load regulation:

   _____ .

8. Reconnect the three 330-Ω resistors.

9. Increase the input voltage to +16.5 V. Record the output voltage as the high-line output:

   $V_{HL} = $ _____ .

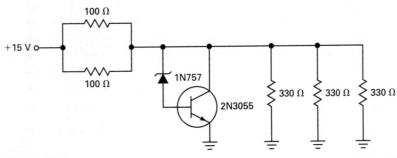

Figure 59-1

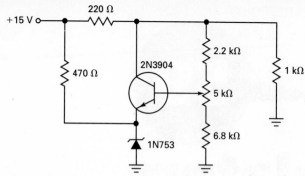

Figure 59-2

10. Reduce the input voltage to +13.5 V. Record the output voltage as the low-line output:

$V_{LL} =$ _____ .

11. Watch the DMM reading of output voltage while you continue to reduce the input voltage. Notice how the regulator stops regulating when there is insufficient input voltage.

12. Calculate and record the line regulation:

_____ .

ADJUSTABLE SHUNT REGULATOR

13. Connect the circuit of Fig. 59-2.
14. Use a DMM to measure and record the zener voltage:

$V_Z =$ _____ .

15. Measure the load voltage across the 1-kΩ resistor. Vary the 5-kΩ potentiometer in both directions to record the minimum and maximum load voltages:

$V_{L(max)} =$ _____ .

$V_{L(min)} =$ _____ .

16. Adjust the potentiometer to get a load voltage of 10 V. Record this value as the full-load voltage:

$V_{FL} =$ _____ .

17. Remove the 1-kΩ load resistor and record this as the no-load voltage:

$V_{NL} =$ _____ .

18. Calculate and record the load regulation:

_____ .

19. Reconnect the 1-kΩ load resistor.
20. Increase the input voltage to +16.5 V. Record the output voltage as the high-line output:

$V_{HL} =$ _____ .

21. Reduce the input voltage to +13.5 V. Record the output voltage as the low-line output:

$V_{LL} =$ _____ .

22. Watch the DMM reading of output voltage while you continue to reduce the input voltage. Notice how the regulator stops regulating when there is insufficient input voltage.

23. Calculate and record the line regulation:

_____ .

OP-AMP SHUNT REGULATOR

24. Connect the circuit of Fig. 59-3.
25. Use a DMM to measure and record the zener voltage:

$V_Z =$ _____ .

26. Measure the load voltage across the 1-kΩ resistor. Vary the 1-kΩ potentiometer in both directions to record the minimum and maximum load voltages:

$V_{L(max)} =$ _____ .

$V_{L(min)} =$ _____ .

27. Adjust the potentiometer to get a load voltage as close as possible to 10 V. Record this value as the full-load voltage:

$V_{FL} =$ _____ .

28. Remove the 1-kΩ load resistor and record this as the no-load voltage:

$V_{NL} =$ _____ .

29. Calculate and record the load regulation:

_____ .

30. Reconnect the 1-kΩ load resistor.
31. Increase the input voltage to +16.5 V. Record the output voltage as the high-line output:

$V_{HL} =$ _____ .

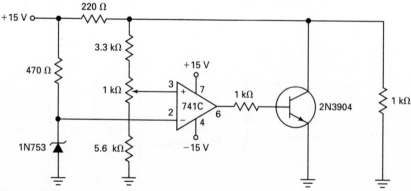

Figure 59-3

276

32. Reduce the input voltage to $+13.5$ V. Record the output voltage as the low-line output:

$V_{LL} = $ _____ .

33. Calculate and record the line regulation: _____ .

COMPUTER (OPTIONAL)

34. Repeat Steps 1 to 33 using EWB or an equivalent circuit simulator. Do not record any new values. But make sure that you get reasonable agreement between the EWB measurements and the values recorded earlier.

35. If you are using the CD-ROM version of this book, click on the Assignments menu and select Chap. 24.

ADDITIONAL WORK (OPTIONAL)

36. Graph the load voltage versus the load resistance for the circuit of Fig. 59-1 using Steps 3 to 6.
37. Graph the load voltage versus the load current for the circuit of Fig. 59-1 using Steps 3 to 6.
38. Have another student insert one of the following troubles into any of the three circuits: open any component or connecting wire. Use only voltage readings of a DMM to troubleshoot.
39. Repeat Step 38 several times until you are confident that you can troubleshoot the circuit for various troubles.

Questions for Experiment 59

1. When the load current increases in Fig. 59-1, the zener current: ( )
   (a) decreases;   (b) increases;   (c) stays the same.
2. In Fig. 59-1, the input current to the shunt regulator is approximately: ( )
   (a) 25 mA;   (b) 50 mA;   (c) 100 mA;   (d) 200 mA.
3. If the 2N3055 has a current gain of 50 in Fig. 59-1, the zener current with no load   ( )
   is:
   (a) 0;   (b) 1 mA;   (c) 2 mA;   (d) 5 mA.
4. When the wiper is moved down in Fig. 59-2, the load voltage: ( )
   (a) decreases;   (b) increases;   (c) stays the same.
5. In Fig. 59-2, the zener current is closest to: ( )
   (a) 20 mA;   (b) 30 mA;   (c) 40 mA;   (d) 50 mA.
6. When the wiper is moved up in Fig. 59-3, the load voltage: ( )
   (a) decreases;   (b) increases;   (c) stays the same.
7. One way to increase the maximum load current in Fig. 59-3 is to use a 2N3055   ( )
   and to decrease the:
   (a) 220 Ω;   (b) 470 Ω;   (c) 2.2 kΩ;   (d) 6.8 kΩ.
8. When load current increases in Fig. 59-3, the zener current: ( )
   (a) decreases;   (b) increases;   (c) stays the same.
9. If the load resistance decreases in Fig. 59-3, the output voltage of the op amp: ( )
   (a) decreases;   (b) increases.
10. Optional. Instructor's question.

Series Regulators

The dc voltage out of a bridge rectifier has a peak-to-peak ripple typically around 10 percent of the unregulated dc voltage. By using this unregulated voltage as the input to a voltage regulator, we can produce a dc output voltage that is almost constant with very small ripple. A voltage regulator uses noninverting voltage feedback. The input or reference voltage comes from a zener diode. This zener voltage is amplified by the closed-loop voltage gain of the regulator. The result is a larger dc output voltage with the same temperature coefficient as the zener diode. Most voltage regulators include current limiting to prevent an accidental short across the load terminals from destroying the pass transistor or diodes in the unregulated supply.

Required Reading

Chapter 24 (Secs. 24-1 to 24-3) of *Electronic Principles*, 6th ed.

Equipment

- 1 power supply: adjustable from 0 to 15 V
- 11 ½-W resistors: 100 Ω, 220 Ω, 330 Ω, 470 Ω, two 680 Ω, 1 kΩ, two 2.2 kΩ, 4.7 kΩ, 10 kΩ
- 1 zener diode: 1N753
- 3 transistors: 2N3904
- 1 capacitor: 0.1 μF
- 1 VOM (analog or digital multimeter)

Procedure

MINIMUM AND MAXIMUM LOAD VOLTAGE

1. In Fig. 60-1, what is the approximate voltage across the zener diode? Record in Table 60-1. (*Note:* A bypass capacitor of 0.1 μF is used to prevent oscillations.)
2. When R_5 is varied, the load voltage changes in Fig. 60-1. Calculate and record the minimum and maximum load voltage.
3. Connect the circuit.
4. Adjust the dc input voltage V_S to +15 V. Measure and record the zener voltage.
5. Adjust R_5 to get the minimum load voltage. Measure and record $V_{L(min)}$.
6. Adjust R_5 to get maximum load voltage. Measure and record $V_{L(max)}$.

LOAD REGULATION

7. Adjust R_5 to get a load voltage of 10 V. Record this as the no-load voltage in Table 60-2.
8. Connect a load resistance of 1 kΩ. Measure the load voltage. Record this as the full-load voltage in Table 60-2.
9. Calculate and record the percent load regulation.

LINE REGULATION

10. Measure the load voltage. Record this as $V_{L(max)}$ under source regulation in Table 60-2.
11. Decrease the input voltage from +15 to +12 V. This represents a line change of approximately 20 percent. Measure and record the load voltage as $V_{L(min)}$.
12. Return the input voltage to +15 V. Calculate and record the percent source regulation.

CURRENT LIMITING

13. Assume that the load voltage is 10 V and that Q_3 turns on when V_{BE} is 0.7 V. Notice that the R_4-R_5-R_6 voltage divider has some current through it. Calculate the load current where current limiting begins in Fig. 60-1. Record this at the top of Table 60-3.
14. Connect an R_L of 10 kΩ. Adjust the load voltage to 10 V. Then measure and record the load voltage for each load resistance listed in Table 60-3.
15. Connect a load resistance of 1 kΩ. Short the load terminals and notice how the load voltage goes to zero. Remove the short from the load terminals and notice how the load voltage returns to normal.

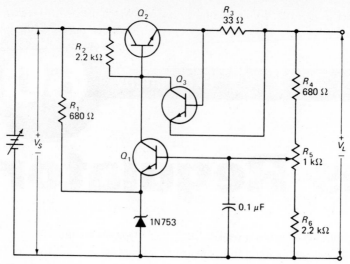

Figure 60-1

16. Use the VOM as an ammeter. Connect the VOM directly across the load terminals. It now measures the load current with a shorted load. This reading should be in the vicinity of your calculated I_{SL} at the top of Table 60-3.

TROUBLESHOOTING

17. For each trouble listed in Table 60-4, estimate and record the load voltage.
18. Insert each trouble into the circuit. Measure and record the load voltage.

CRITICAL THINKING

19. Select a value for R_4 to get a theoretical load voltage of approximately 9 to 13.2 V.
20. Connect the circuit with your value of R_4.
21. Measure the minimum and maximum load voltage. Record all quantities listed in Table 60-5.

COMPUTER (OPTIONAL)

22. Repeat Steps 1 to 21 using EWB or an equivalent circuit simulator. Do not record any new values. But make sure that you get reasonable agreement between the EWB measurements and the values recorded earlier.
23. If you are using the CD-ROM version of this book, click on the Assignments menu and select Chap. 24.

ADDITIONAL WORK (OPTIONAL)

24. Connect the series regulator of Fig. 60-2.
25. Use a DMM to measure and record the zener voltage:

$$V_Z = \underline{\hspace{2cm}} .$$

26. Measure the load voltage across the 1-kΩ resistor. Vary the 1-kΩ potentiometer in both directions to record the minimum and maximum load voltages:

$$V_{L(max)} = \underline{\hspace{2cm}} .$$

$$V_{L(min)} = \underline{\hspace{2cm}} .$$

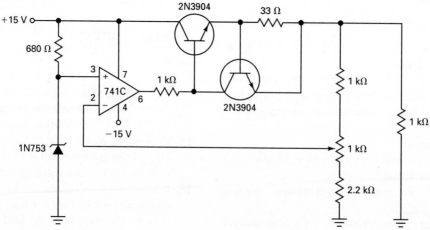

Figure 60-2

282

27. Adjust the potentiometer to get a load voltage of 10 V. Record this value as the full-load voltage:

$V_{FL} = $ _____ .

28. Remove the 1-kΩ load resistor and record this as the no-load voltage:

$V_{NL} = $ _____ .

29. Calculate and record the load regulation:

_____ .

30. Reconnect the 1-kΩ load resistor.

31. Increase the input voltage to +16.5 V. Record the output voltage as the high-line output:

$V_{HL} = $ _____ .

32. Reduce the input voltage to +13.5 V. Record the output voltage as the low-line output:

$V_{LL} = $ _____ .

33. Calculate and record the load regulation:

_____ .

34. Watch the DMM reading of output voltage while you continue to reduce the input voltage. Notice how the regulator stops regulating when there is insufficient input voltage.

35. Adjust the input voltage to 15 V. Connect an ammeter across the load to produce the shorted-load current. Record the value:

$I_{SL} = $ _____ .

36. What is the power dissipation in the pass transistor when a short is across the load?

$P_D = $ _____ .

37. Briefly describe how to modify the circuit shown in Fig. 60-2 to supply a regulated 10 V across a 10-Ω load resistor:

Data for Experiment 60

TABLE 60-1. MINIMUM AND MAXIMUM LOAD VOLTAGE

Calculated	Measured
$V_Z =$ _____	$V_Z =$ _____
$V_{L(min)} =$ _____	$V_{L(min)} =$ _____
$V_{L(max)} =$ _____	$V_{L(max)} =$ _____

TABLE 60-2. REGULATION

Load Regulation	Source Regulation
$V_{NL} =$ _____	$V_{L(max)} =$ _____
$V_{FL} =$ _____	$V_{L(min)} =$ _____
%LR = _____	%SR = _____

TABLE 60-3. CURRENT LIMITING: $I_{SL} =$ _____

R_L	V_L
10 kΩ	
4.7 kΩ	
1 kΩ	
470 Ω	
330 Ω	
220 Ω	
100 Ω	
0	

TABLE 60-4. TROUBLESHOOTING

Trouble	Estimated V_{out}	Measured V_{out}
R_2 open		
Zener open		
Zener short		
Q_1 open		

TABLE 60-5. CRITICAL THINKING

$R_4 =$	
$V_{L(min)} =$	
$V_{L(max)} =$	

Questions for Experiment 60

1. The zener voltage of Fig. 60-1 is approximately: ( )
   (a) 5 V;   (b) 6.2 V;   (c) 7.5 V;   (d) 15 V.
2. Theoretically, the maximum regulated load voltage of Fig. 60-1 is approximately: ( )
   (a) 6.2 V;   (b) 8.37 V;   (c) 12.2 V;   (d) 15 V.
3. Current limiting of Fig. 60-1 starts near: ( )
   (a) 1 mA;   (b) 2.25 mA;   (c) 12.5 mA;   (d) 18.6 mA.
4. The data of Table 60-2 show that load voltage is: ( )
   (a) dependent on load current;   (b) proportional to source voltage;
   (c) almost constant;   (d) low.
5. When V_L is +10 V and R_L is 1 kΩ in Fig. 60-1, the power dissipation in the pass ( )
   transistor is approximately:
   (a) 50 mW;   (b) 100 mW;   (c) 200 mW;   (d) 279 mW.
6. Briefly explain how the voltage regulator of Fig. 60-1 works.

7. Assume the load terminals are shorted in Fig. 60-1. If the source voltage is +15 V, what is the power dissipation in the pass transistor?

TROUBLESHOOTING

8. Why does the load voltage approach the source voltage when the zener diode opens in Fig. 60-1?

CRITICAL THINKING

9. Suppose we want current limiting to start at approximately 100 mA in Fig. 60-1. What changes are necessary?

10. Optional. Instructor's question.

Three-Terminal IC Regulators

The LM78XX series is typical of the IC voltage regulators currently available. These three-terminal regulators are the ultimate in simplicity. And they are virtually indestructible because of the thermal shutdown discussed in your textbook. In this experiment you will connect an LM7805 as a voltage regulator and a current regulator.

Required Reading

Chapter 24 (Sec. 24-4) of *Electronic Principles*, 6th ed.

Equipment

1 audio generator
1 power supply: adjustable from 0 to 15 V
1 voltage regulator: LM7805
6 ½-W resistors: 10 Ω, 22 Ω, 33 Ω, 47 Ω, 68 Ω, 150 Ω
2 capacitors: 0.1 μF, 0.22 μF
1 VOM (analog or digital multimeter)
1 oscilloscope

Procedure

VOLTAGE REGULATOR

1. In Fig. 61-1, estimate and record the output voltage for each input voltage listed in Table 61-1.
2. Connect the circuit.
3. Measure and record the output voltage for each input voltage listed in Table 61-1.

RIPPLE REJECTION

4. In Fig. 61-2, the ac source in series with the dc source simulates ripple superimposed on dc voltage. The data sheet of an LM7805 (see the Appendix) lists the following ripple rejection: minimum is 62 dB and typical is 80 dB. Calculate the peak-to-peak ac output for the minimum and typical ripple rejection. Record your answers in Table 61-2.
5. Connect the circuit of Fig. 61-2.
6. Look at the ac voltage at the input to the regulator. Adjust the signal source to get 2 V pp at 120 Hz.
7. Use the most sensitive ac ranges of the oscilloscope to look at the output ripple. Measure and record this ac output voltage. Then calculate and record the ripple rejection ratio in decibels.

ADJUSTABLE VOLTAGE REGULATOR AND CURRENT REGULATOR

8. The circuit of Fig. 61-3 can function either as a voltage regulator if you use the output voltage or as a current regulator if R_2 is the load resistor. Calculate and record V_{out} and I_{out} for each value of R_2 listed in Table 61-3.

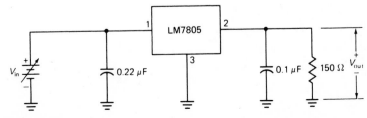

Figure 61-1

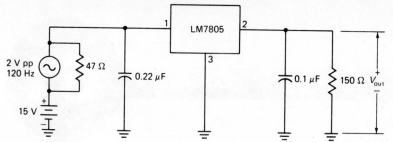

Figure 61-2

9. Connect the circuit with an R_2 of 10 Ω. Measure and record V_{out} and I_{out}.

10. Repeat Step 9 for the other values of R_2.

TROUBLESHOOTING

11. Assume that the circuit of Fig. 61-3 has an R_2 of 68 Ω. For each trouble in Table 61-4, estimate and record the dc output voltage.

12. Connect the circuit with an R_2 of 68 Ω. Insert each trouble. Measure and record the dc output voltage.

CRITICAL THINKING

13. Select a value of R_2 in Fig. 61-3 to produce an output voltage of approximately 9 V.

14. Insert your value of R_2. Measure the output voltage. Record R_2 and V_{out} in Table 61-5.

APPLICATION (OPTIONAL)—REGULATED POWER SUPPLY

15. In Fig. 61-4, the secondary voltage is 12.6 VAC. Calculate the ideal dc input voltage and ripple to the LM7805:

$V_{dc} = $ _____ .

$V_{rip} = $ _____ .

16. Connect the circuit of Fig. 61-4 with the same power transformer used in Experiment 8. Also, use 1N4001s.

17. Measure and record the dc input voltage and ripple to the LM7805:

$V_{dc} = $ _____ .

$V_{rip} = $ _____ .

18. Measure and record the dc output voltage and ripple of the LM7805:

$V_{dc} = $ _____ .

$V_{rip} = $ _____ .

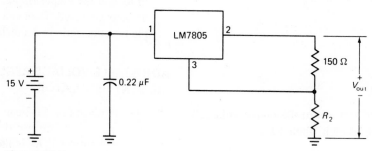

Figure 61-3

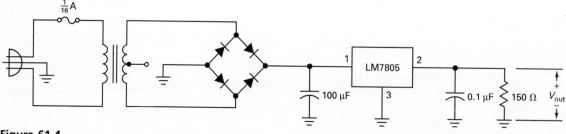

Figure 61-4

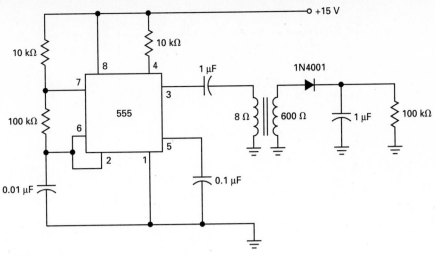

Figure 61-5

19. Compare the input and output values of Steps 17 and 18 and then discuss why these values occur:

24. Use the oscilloscope on ac coupling to measure the peak-to-peak ripple across the load resistor:

_____ .

25. Explain why the final output voltage is a dc voltage larger than the supply voltage:

DC-TO-DC CONVERTER

20. Section 24-6 of the textbook discusses dc-to-dc converters. In this demonstration you will build the dc-to-dc converter of Fig. 61-5. An astable 555 multivibrator drives the 8-Ω side of the audio transformer used in Experiment 26. The 600-Ω side drives a half-wave rectifier with a capacitor-input filter.

21. Connect the dc-to-dc converter of Fig. 61-5.

22. Look at the waveform at pin 3. Record the peak-to-peak value:

_____ .

23. Measure the dc voltage across the final 100-kΩ resistor with a DMM. Record the value:

_____ .

COMPUTER (OPTIONAL)

26. Repeat Steps 1 to 14 using EWB or an equivalent circuit simulator. Do not record any new values. But make sure that you get reasonable agreement between the EWB measurements and the values recorded earlier.

27. If you are using the CD-ROM version of this book, click on the Assignments menu and select Chap. 24.

Data for Experiment 61

TABLE 61-5. VOLTAGE REGULATOR

V_{in}	Estimated V_{out}	Measured V_{out}
1 V		
5 V		
10 V		
11 V		
12 V		
13 V		
14 V		
15 V		

TABLE 61-2. RIPPLE REJECTION

Calculated V_{rip} (minimum rejection) =

Calculated V_{rip} (typical rejection) =

Measured V_{rip} =

Measured ripple rejection =

TABLE 61-3. VOLTAGE AND CURRENT REGULATION

R_2	Calculated		Measured	
	V_{out}	I_{out}	V_{out}	I_{out}
10 Ω				
22 Ω				
33 Ω				
47 Ω				
68 Ω				

TABLE 61-4. TROUBLESHOOTING

Trouble	Estimated V_{out}	Measured V_{out}
R_1 short		
R_1 open		
R_2 short		
R_2 open		

TABLE 61-5. CRITICAL THINKING

$R_2 =$ _____

$V_{out} =$ _____

Questions for Experiment 61

1. When the input voltage of Fig. 61-1 is greater than 10 V, the output voltage is ( )
   approximately:
   (a) constant;   (b) 5 V;   (c) regulated;   (d) all of the foregoing.
2. In Table 61-2, the typical output ripple is approximately: ( )
   (a) 0.2 mV;   (b) 1 mV;   (c) 1.59 mV;   (d) 10 mV.
3. If I_Q is 8 mA in Fig. 61-3, the calculated I_{out} is approximately: ( )
   (a) 8 mA;   (b) 23.3 mA;   (c) 41.3 mA;   (d) 100 mA.
4. When R_2 is 68 Ω in Fig. 61-3, the calculated V_{out} is approximately: ( )
   (a) 5.43 V;   (b) 6.43 V;   (c) 7.04 V;   (d) 7.94 V.
5. The measured current in Table 61-3 indicates that the regulator circuit can func- ( )
   tion as a:
   (a) current source;   (b) voltage source;   (c) ripple generator;
   (d) amplifier.
6. Briefly explain why the data sheet of an LM340-5 indicates that the input voltage must be
   at least 7 V.

7. Why are bypass capacitors used with an IC regulator?

TROUBLESHOOTING

8. What output voltage did you get when R_1 was open? Why did you get this voltage?

CRITICAL THINKING

9. What value did you use for R_2 to get an output of 9 V? How did you arrive at this value?

10. Optional. Instructor's question.

Appendix

Parts and Equipment

RESISTORS (ALL ½ W UNLESS OTHERWISE SPECIFIED)

Quantity	Description
1	10 Ω
2	22 Ω
1	33 Ω
1	47 Ω
1	51 Ω
2	68 Ω
2	100 Ω
1	100 Ω, 1 W
2	150 Ω
1	180 Ω
2	220 Ω
2	270 Ω
2	270 Ω, 1 W
3	330 Ω
4	470 Ω
1	470 Ω, 1 W
1	560 Ω
2	680 Ω
2	820 Ω
1	910 Ω
4	1 kΩ
1	1.1 kΩ
1	1.2 kΩ
2	1.5 kΩ
1	1.8 kΩ
1	2 kΩ
2	2.2 kΩ
1	2.7 kΩ
1	3.3 kΩ
3	3.9 kΩ
2	4.7 kΩ
2	6.8 kΩ
1	8.2 kΩ
3	10 kΩ
2	15 kΩ
1	18 kΩ
2	22 kΩ
1	27 kΩ
3	33 kΩ
4	39 kΩ
2	47 kΩ
1	68 kΩ
2	100 kΩ
2	220 kΩ
1	270 kΩ
1	330 kΩ
1	470 kΩ
1	680 kΩ
2	1 MΩ

POTENTIOMETERS

Quantity	Description
1	1 kΩ
1	5 kΩ
2	10 kΩ
1	50 kΩ

CAPACITORS

Quantity	Description
2	100 pF
2	220 pF
1	470 pF
2	0.001 μF
3	0.002 μF
1	0.003 μF
2	0.0047 μF
2	0.01 μF
1	0.02 μF
2	0.022 μF
2	0.047 μF
1	0.068 μF
2	0.1 μF
1	0.22 μF
1	0.33 μF

CAPACITORS (cont.)

Quantity	Description
4	0.47 μF
3	1 μF
2	10 μF
2	47 μF
1	100 μF
2	470 μF

DIODES

Quantity	Description
3	1N753 (6.2-V zener)
1	1N757 (9-V zener)
4	1N4001 (rectifier)
3	1N4148 (small-signal)
1	L53RD (red LED)
1	L53GD (green LED)
1	RS276-1067 (SCR)

TRANSISTORS

Quantity	Description
1	2N3055 (power *npn*)
3	2N3904 (small-signal *npn*)
3	2N3906 (small-signal *pnp*)
1	IFR510 (power FET)
3	MPF102 (*n*-channel JFET)

INTEGRATED CIRCUITS

Quantity	Description
1	LM318C
3	LM741C
1	LM7805
2	NE555

MISCELLANEOUS PARTS

Quantity	Description
1	Inductor
1	TIL312 (7-segment)
1	Optocoupler: 4N26
2	SPST switch
1	F-25X (fused power transformer with 12.6 V ac center-tapped secondary)

OPTIONAL PARTS FOR APPLICATIONS

(*Note:* All are Radio Shack components)

Quantity	Description
1	RS276-1657 (Photocells)
1	RS270-090C (Microphone element)
1	RS273-026 (Magnetic buzzer)
1	RS273-256 (Motor)
1	RS273-223 (Motor)
1	RS276-145A (Phototransistor)
1	RS276-2072A (Power FET)
1	RS276-1067 (SCR)

Data for Radio Shack Devices

Cadmium-sulfide photocells: RS276-1657
    Assorted photoresistors
    Light resistance of some units is less than 100 Ω
    Dark resistance of some units is more than 1 MΩ
    Dark-light resistance ratio of any unit is at least 100:1
Condenser microphone element: RS270-090C
    Supply voltage: 1 to 10 V dc
    Nominal supply: 4.5 V dc
    Current drain: 0.3 mA
    Signal/noise: 60 dB
    Output impedance: 1 kΩ
Magnetic buzzer: RS273-026
    Voltage range: 6 to 16 V dc
    Current drain: 10 mA at 12 V dc
    Frequency: 2.9 to 3.9 kHz
    Polarized: Red lead is positive, and black lead is
      negative
Motor: RS273-256
    Voltage range: 9 to 18 V dc
    No-load current drain: 0.3 to 0.4 A
    Full-load current drain: 1.98 A
    RPM: 24,000 no load, 18,000 full load
    Nonpolarized
Motor: RS273-223
    Voltage range: 1.5 to 3 V dc
    No-load current drain: 0.2 to 0.3 A
    Full-load current drain: 0.98 A
    RPM: 5700-11,600 no load, 4100-8300 full load
    Nonpolarized

Phototransistor: RS276-145A
    $V_{CE(max)}$: 30 V
    $V_{CE(sat)}$: 0.4 V
    Dark current: 100 nA
    Light current: 20 mA
    Risetime: 5 μs
    Fall time: 5 μs
Power FET: RS276-2072A
    Absolute maximum ratings
      Drain-source voltage: 100 V
      Drain-gate voltage: 100 V
      Continuous drain current: 4 A
      Pulsed drain current: 16 A
      Gate-source voltage: 20 V
      Maximum power dissipation: 20 W
Electrical specifications
    Gate threshold voltage: 2 to 4 V
    On-state resistance: 0.54 Ω
    Tranconductance: 1 mho (S)
    Input capacitance: 150 pF
    Output capacitance: 100 pF
Silicon-controlled rectifier: RS276-1067
    Maximum ratings
      Gate trigger current: 25 mA
      Holding current: 25 mA
      On-state current: 6 A
      Peak reverse anode voltage: 200 V
      Peak reverse gate voltage: 10 V
      Peak gate power dissipation: 5 W

National
Semiconductor™

1N746A - 1N759A Series Half Watt Zeners

Absolute Maximum Ratings*   TA = 25°C unless otherwise noted

Tolerance: A = 5%

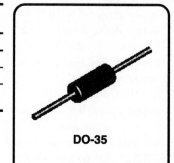

DO-35

Parameter	Value	Units
Storage Temperature Range	-65 to +200	°C
Maximum Junction Operating Temperature	+175	°C
Lead Temperature (1/16" from case for 10 seconds)	+230	°C
Total Device Dissipation	500	mW
Derate above 25°C	3.33	mW/°C

*These ratings are limiting values above which the serviceability of the diode may be impaired.

<u>NOTES:</u>
1) These ratings are based on a maximum junction temperature of 200 degrees C.
2) These are steady state limits. The factory should be consulted on applications involving pulsed or low duty cycle operations.

Electrical Characteristics   TA = 25°C unless otherwise noted

Device	V_Z (V)	Z_Z (Ω)	@ I_{ZT} (mA)	I_{R1} (µA)	@ V_R (V)	I_{R2} (µA)	@ V_R TA=150°C (V)	T_C (%/°C)	I_{ZM}* (mA)
1N746A	3.3	28	20	10	1.0	30	1.0	- 0.070	110
1N747A	3.6	24	20	10	1.0	30	1.0	- 0.065	100
1N748A	3.9	23	20	10	1.0	30	1.0	- 0.060	95
1N749A	4.3	22	20	2.0	1.0	30	1.0	+/- 0.055	85
1N750A	4.7	19	20	2.0	1.0	30	1.0	+/- 0.030	75
1N751A	5.1	17	20	1.0	1.0	20	1.0	+/- 0.030	70
1N752A	5.6	11	20	1.0	1.0	20	1.0	+ 0.038	65
1N753A	6.2	7.0	20	0.1	1.0	20	1.0	+ 0.045	60
1N754A	6.8	5.0	20	0.1	1.0	20	1.0	+ 0.050	55
1N755A	7.5	6.0	20	0.1	1.0	20	1.0	+ 0.058	50
1N756A	8.2	8.0	20	0.1	1.0	20	1.0	+ 0.062	45
1N757A	9.1	10	20	0.1	1.0	20	1.0	+ 0.068	40
1N758A	10	17	20	0.1	1.0	20	1.0	+ 0.075	35
1N759A	12	30	20	0.1	1.0	20	1.0	+ 0.077	38

*IZM (Maximum Zener Current Rating) Values shown are based on the JEDEC rating of 400 milliwatts. Where the actual zener voltage (VZ) is known at the operating point, the maximum zener current may be increased and is limited by the derating curve.

Axial Lead
Standard Recovery Rectifiers

This data sheet provides information on subminiature size, axial lead mounted rectifiers for general–purpose low–power applications.

Mechanical Characteristics

- Case: Epoxy, Molded
- Weight: 0.4 gram (approximately)
- Finish: All External Surfaces Corrosion Resistant and Terminal Leads are Readily Solderable
- Lead and Mounting Surface Temperature for Soldering Purposes: 220°C Max. for 10 Seconds, 1/16″ from case
- Shipped in plastic bags, 1000 per bag.
- Available Tape and Reeled, 5000 per reel, by adding a "RL" suffix to the part number
- Polarity: Cathode Indicated by Polarity Band
- Marking: 1N4001, 1N4002, 1N4003, 1N4004, 1N4005, 1N4006, 1N4007

**1N4001
thru
1N4007**

1N4004 and 1N4007 are
Motorola Preferred Devices

**LEAD MOUNTED
RECTIFIERS
50–1000 VOLTS
DIFFUSED JUNCTION**

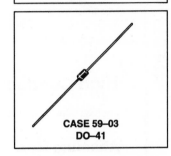

**CASE 59–03
DO–41**

MAXIMUM RATINGS

Rating	Symbol	1N4001	1N4002	1N4003	1N4004	1N4005	1N4006	1N4007	Unit
*Peak Repetitive Reverse Voltage Working Peak Reverse Voltage DC Blocking Voltage	V_{RRM} V_{RWM} V_R	50	100	200	400	600	800	1000	Volts
*Non–Repetitive Peak Reverse Voltage (halfwave, single phase, 60 Hz)	V_{RSM}	60	120	240	480	720	1000	1200	Volts
*RMS Reverse Voltage	$V_{R(RMS)}$	35	70	140	280	420	560	700	Volts
*Average Rectified Forward Current (single phase, resistive load, 60 Hz, see Figure 8, T_A = 75°C)	I_O	1.0							Amp
*Non–Repetitive Peak Surge Current (surge applied at rated load conditions, see Figure 2)	I_{FSM}	30 (for 1 cycle)							Amp
Operating and Storage Junction Temperature Range	T_J T_{stg}	– 65 to +175							°C

ELECTRICAL CHARACTERISTICS*

Rating	Symbol	Typ	Max	Unit
Maximum Instantaneous Forward Voltage Drop (i_F = 1.0 Amp, T_J = 25°C) Figure 1	v_F	0.93	1.1	Volts
Maximum Full–Cycle Average Forward Voltage Drop (I_O = 1.0 Amp, T_L = 75°C, 1 inch leads)	$V_{F(AV)}$	—	0.8	Volts
Maximum Reverse Current (rated dc voltage) (T_J = 25°C) (T_J = 100°C)	I_R	0.05 1.0	10 50	μA
Maximum Full–Cycle Average Reverse Current (I_O = 1.0 Amp, T_L = 75°C, 1 inch leads)	$I_{R(AV)}$	—	30	μA

*Indicates JEDEC Registered Data

Preferred devices are Motorola recommended choices for future use and best overall value.

Rev 5

MOTOROLA

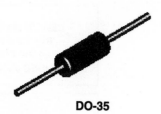

National Semiconductor™

Discrete POWER & Signal Technologies

1N/FDLL 914/A/B / 916/A/B / 4148 / 4448

DO-35

LL-34

THE PLACEMENT OF THE EXPANSION GAP HAS NO RELATIONSHIP TO THE LOCATION OF THE CATHODE TERMINAL

COLOR BAND MARKING		
DEVICE	**1ST BAND**	**2ND BAND**
FDLL914	BLACK	BROWN
FDLL914A	BLACK	GRAY
FDLL914B	BROWN	BLACK
FDLL916	BLACK	RED
FDLL916A	BLACK	WHITE
FDLL916B	BROWN	BROWN
FDLL4148	BLACK	BROWN
FDLL4448	BROWN	BLACK

High Conductance Fast Diode

Sourced from Process D3.

Absolute Maximum Ratings*   TA = 25°C unless otherwise noted

Symbol	Parameter	Value	Units
W_{IV}	Working Inverse Voltage	75	V
I_O	Average Rectified Current	200	mA
I_F	DC Forward Current	300	mA
i_f	Recurrent Peak Forward Current	400	mA

$i_{f(surge)}$	Peak Forward Surge Current
	    Pulse width = 1.0 second
	    Pulse width = 1.0 microsecond
		
		1.0
		4.0
			
			A
			A

T_{stg}	Storage Temperature Range	-65 to +200	°C
T_J	Operating Junction Temperature	175	°C

*These ratings are limiting values above which the serviceability of any semiconductor device may be impaired.

<u>NOTES</u>:
1) These ratings are based on a maximum junction temperature of 200 degrees C.
2) These are steady state limits. The factory should be consulted on applications involving pulsed or low duty cycle operations.

Thermal Characteristics   TA = 25°C unless otherwise noted

Symbol	Characteristic	Max	Units
		1N/FDLL 914/A/B / 4148 / 4448	
P_D	Total Device Dissipation Derate above 25°C	500 3.33	mW mW/°C
$R_{\theta JA}$	Thermal Resistance, Junction to Ambient	300	°C/W

Electrical Characteristics   TA = 25°C unless otherwise noted

Symbol	Parameter		Test Conditions	Min	Max	Units
B_V	Breakdown Voltage		$I_R = 100\ \mu A$	100		V
			$I_R = 5.0\ \mu A$	75		V
I_R	Reverse Current		$V_R = 20\ V$		25	nA
			$V_R = 20\ V,\ T_A = 150°C$		50	μA
			$V_R = 75\ V$		5.0	μA
V_F	Forward Voltage	1N914B / 4448	$I_F = 5.0\ mA$	620	720	mV
		1N916B	$I_F = 5.0\ mA$	630	730	mV
		1N914 / 916 / 4148	$I_F = 10\ mA$		1.0	V
		1N914A / 916A	$I_F = 20\ mA$		1.0	V
		1N916B	$I_F = 30\ mA$		1.0	V
		1N914B / 4448	$I_F = 100\ mA$		1.0	V
C_O	Diode Capacitance					
		1N916/A/B / 4448	$V_R = 0,\ f = 1.0\ MHz$		2.0	pF
		1N914/A/B / 4148	$V_R = 0,\ f = 1.0\ MHz$		4.0	pF
T_{RR}	Reverse Recovery Time		$I_F = 10\ mA,\ V_R = 6.0\ V\ (60\ mA),$ $I_{rr} = 1.0\ mA,\ R_L = 100\ \Omega$		4.0	nS

Typical Characteristics

REVERSE VOLTAGE vs REVERSE CURRENT
BV - 1.0 to 100 uA

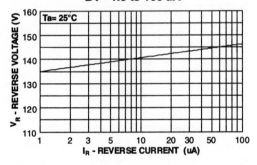

REVERSE CURRENT vs REVERSE VOLTAGE
IR - 10 to 100 V

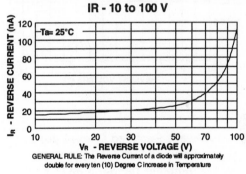

GENERAL RULE: The Reverse Current of a diode will approximately
double for every ten (10) Degree C increase in Temperature

FORWARD VOLTAGE vs FORWARD CURRENT
VF - 1 to 100 uA

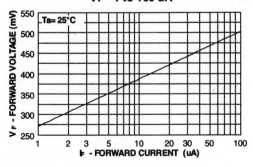

FORWARD VOLTAGE vs FORWARD CURRENT
VF - 0.1 to 100 mA

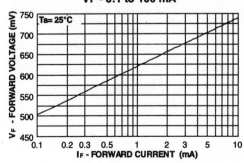

High Conductance Fast Diode
(continued)

Typical Characteristics (continued)

FORWARD VOLTAGE vs FORWARD CURRENT
VF - 10 to 800 mA

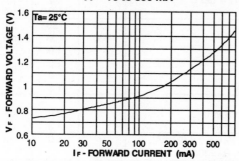

VF - 0.01 - 20 mA (-40 to +65 Deg C)
FORWARD VOLTAGE vs
AMBIENT TEMPERATURE

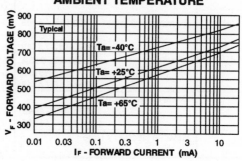

CAPACITANCE vs REVERSE VOLTAGE
VR = 0.0 to 15 V

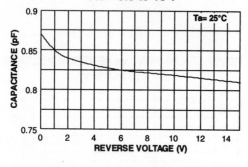

REVERSE RECOVERY TIME vs
REVERSE CURRENT

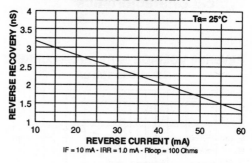

IF = 10 mA - IRR = 1.0 mA - Rloop = 100 Ohms

Average Rectified Current (Io) &
Forward Current (I$_F$) versus
Ambient Temperature (T$_A$)

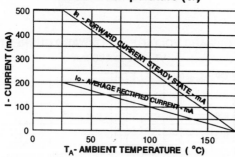

POWER DERATING CURVE

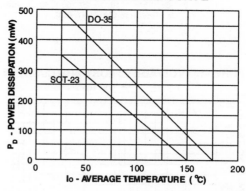

MOTOROLA
SEMICONDUCTOR TECHNICAL DATA

Complementary Silicon Power Transistors

. . . designed for general–purpose switching and amplifier applications.

- DC Current Gain — h_{FE} = 20–70 @ I_C = 4 Adc
- Collector–Emitter Saturation Voltage —
  $V_{CE(sat)}$ = 1.1 Vdc (Max) @ I_C = 4 Adc
- Excellent Safe Operating Area

NPN
2N3055*
PNP
MJ2955 *

*Motorola Preferred Device

**15 AMPERE
POWER TRANSISTORS
COMPLEMENTARY
SILICON
60 VOLTS
115 WATTS**

**CASE 1–07
TO–204AA
(TO–3)**

MAXIMUM RATINGS

Rating	Symbol	Value	Unit
Collector–Emitter Voltage	V_{CEO}	60	Vdc
Collector–Emitter Voltage	V_{CER}	70	Vdc
Collector–Base Voltage	V_{CB}	100	Vdc
Emitter–Base Voltage	V_{EB}	7	Vdc
Collector Current — Continuous	I_C	15	Adc
Base Current	I_B	7	Adc
Total Power Dissipation @ T_C = 25°C Derate above 25°C	P_D	115 0.657	Watts W/°C
Operating and Storage Junction Temperature Range	T_J, T_{stg}	−65 to +200	°C

THERMAL CHARACTERISTICS

Characteristic	Symbol	Max	Unit
Thermal Resistance, Junction to Case	$R_{\theta JC}$	1.52	°C/W

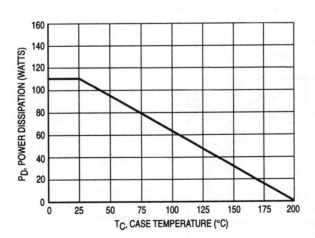

Figure 1. Power Derating

Preferred devices are Motorola recommended choices for future use and best overall value.

Ⓜ **MOTOROLA**

ELECTRICAL CHARACTERISTICS (T_C = 25°C unless otherwise noted)

Characteristic	Symbol	Min	Max	Unit
*OFF CHARACTERISTICS				
Collector–Emitter Sustaining Voltage (1) (I_C = 200 mAdc, I_B = 0)	$V_{CEO(sus)}$	60	—	Vdc
Collector–Emitter Sustaining Voltage (1) (I_C = 200 mAdc, R_{BE} = 100 Ohms)	$V_{CER(sus)}$	70	—	Vdc
Collector Cutoff Current (V_{CE} = 30 Vdc, I_B = 0)	I_{CEO}	—	0.7	mAdc
Collector Cutoff Current (V_{CE} = 100 Vdc, $V_{BE(off)}$ = 1.5 Vdc) (V_{CE} = 100 Vdc, $V_{BE(off)}$ = 1.5 Vdc, T_C = 150°C)	I_{CEX}	— —	1.0 5.0	mAdc
Emitter Cutoff Current (V_{BE} = 7.0 Vdc, I_C = 0)	I_{EBO}	—	5.0	mAdc
*ON CHARACTERISTICS (1)				
DC Current Gain (I_C = 4.0 Adc, V_{CE} = 4.0 Vdc) (I_C = 10 Adc, V_{CE} = 4.0 Vdc)	h_{FE}	20 5.0	70 —	—
Collector–Emitter Saturation Voltage (I_C = 4.0 Adc, I_B = 400 mAdc) (I_C = 10 Adc, I_B = 3.3 Adc)	$V_{CE(sat)}$	—	1.1 3.0	Vdc
Base–Emitter On Voltage (I_C = 4.0 Adc, V_{CE} = 4.0 Vdc)	$V_{BE(on)}$	—	1.5	Vdc
SECOND BREAKDOWN				
Second Breakdown Collector Current with Base Forward Biased (V_{CE} = 40 Vdc, t = 1.0 s, Nonrepetitive)	$I_{s/b}$	2.87	—	Adc
DYNAMIC CHARACTERISTICS				
Current Gain — Bandwidth Product (I_C = 0.5 Adc, V_{CE} = 10 Vdc, f = 1.0 MHz)	f_T	2.5	—	MHz
*Small–Signal Current Gain (I_C = 1.0 Adc, V_{CE} = 4.0 Vdc, f = 1.0 kHz)	h_{fe}	15	120	—
*Small–Signal Current Gain Cutoff Frequency (V_{CE} = 4.0 Vdc, I_C = 1.0 Adc, f = 1.0 kHz)	f_{hfe}	10	—	kHz

* Indicates Within JEDEC Registration. (2N3055)
(1) Pulse Test: Pulse Width ≤ 300 µs, Duty Cycle ≤ 2.0%.

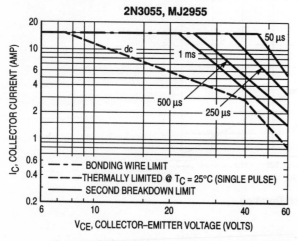

2N3055, MJ2955

Figure 2. Active Region Safe Operating Area

There are two limitations on the power handling ability of a transistor: average junction temperature and second breakdown. Safe operating area curves indicate I_C – V_{CE} limits of the transistor that must be observed for reliable operation; i.e., the transistor must not be subjected to greater dissipation than the curves indicate.

The data of Figure 2 is based on T_C = 25°C; $T_{J(pk)}$ is variable depending on power level. Second breakdown pulse limits are valid for duty cycles to 10% but must be derated for temperature according to Figure 1.

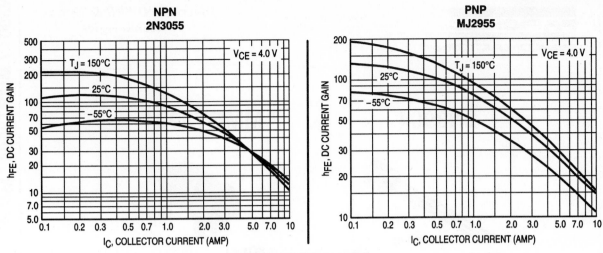

Figure 3. DC Current Gain

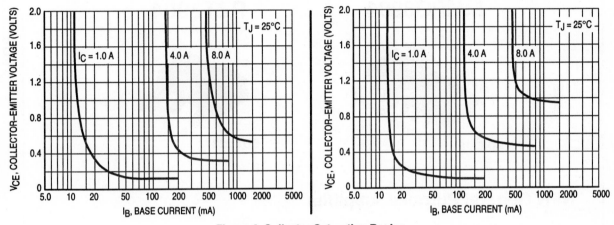

Figure 4. Collector Saturation Region

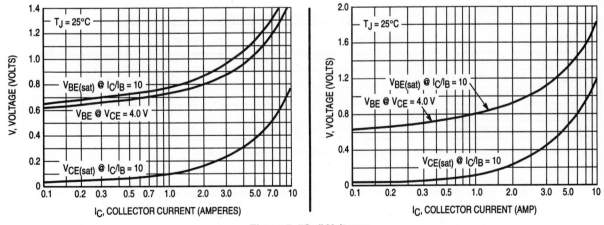

Figure 5. "On" Voltages

Motorola Bipolar Power Transistor Device Data

Discrete POWER & Signal
Technologies

2N3904

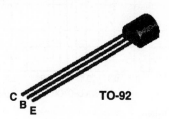

TO-92

MMBT3904

SOT-23
Mark: 1A

MMPQ3904

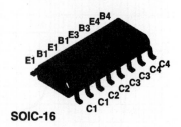

SOIC-16

PZT3904

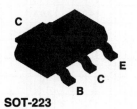

SOT-223

NPN General Purpose Amplifier

This device is designed as a general purpose amplifier and switch.
The useful dynamic range extends to 100 mA as a switch and to
100 MHz as an amplifier. Sourced from Process 23.

Absolute Maximum Ratings* TA = 25°C unless otherwise noted

Symbol	Parameter	Value	Units
V_{CEO}	Collector-Emitter Voltage	40	V
V_{CBO}	Collector-Base Voltage	60	V
V_{EBO}	Emitter-Base Voltage	6.0	V
I_C	Collector Current - Continuous	200	mA
T_J, T_{stg}	Operating and Storage Junction Temperature Range	-55 to +150	°C

*These ratings are limiting values above which the serviceability of any semiconductor device may be impaired.

<u>NOTES</u>:
1) These ratings are based on a maximum junction temperature of 150 degrees C.
2) These are steady state limits. The factory should be consulted on applications involving pulsed or low duty cycle operations.

Electrical Characteristics TA = 25°C unless otherwise noted

Symbol	Parameter	Test Conditions	Min	Max	Units

OFF CHARACTERISTICS

Symbol	Parameter	Test Conditions	Min	Max	Units
$V_{(BR)CEO}$	Collector-Emitter Breakdown Voltage	$I_C = 1.0$ mA, $I_B = 0$	40		V
$V_{(BR)CBO}$	Collector-Base Breakdown Voltage	$I_C = 10$ µA, $I_E = 0$	60		V
$V_{(BR)EBO}$	Emitter-Base Breakdown Voltage	$I_E = 10$ µA, $I_C = 0$	6.0		V
I_{BL}	Base Cutoff Current	$V_{CE} = 30$ V, $V_{EB} = 0$		50	nA
I_{CEX}	Collector Cutoff Current	$V_{CE} = 30$ V, $V_{EB} = 0$		50	nA

ON CHARACTERISTICS*

Symbol	Parameter	Test Conditions	Min	Max	Units
h_{FE}	DC Current Gain	$I_C = 0.1$ mA, $V_{CE} = 1.0$ V	40		
		$I_C = 1.0$ mA, $V_{CE} = 1.0$ V	70		
		$I_C = 10$ mA, $V_{CE} = 1.0$ V	100	300	
		$I_C = 50$ mA, $V_{CE} = 1.0$ V	60		
		$I_C = 100$ mA, $V_{CE} = 1.0$ V	30		
$V_{CE(sat)}$	Collector-Emitter Saturation Voltage	$I_C = 10$ mA, $I_B = 1.0$ mA		0.2	V
		$I_C = 50$ mA, $I_B = 5.0$ mA		0.3	V
$V_{BE(sat)}$	Base-Emitter Saturation Voltage	$I_C = 10$ mA, $I_B = 1.0$ mA	0.65	0.85	V
		$I_C = 50$ mA, $I_B = 5.0$ mA		0.95	V

SMALL SIGNAL CHARACTERISTICS

Symbol	Parameter	Test Conditions	Min	Max	Units
f_T	Current Gain - Bandwidth Product	$I_C = 10$ mA, $V_{CE} = 20$ V, f = 100 MHz	300		MHz
C_{obo}	Output Capacitance	$V_{CB} = 5.0$ V, $I_E = 0$, f = 1.0 MHz		4.0	pF
C_{ibo}	Input Capacitance	$V_{EB} = 0.5$ V, $I_C = 0$, f = 1.0 MHz		8.0	pF
NF	Noise Figure (except MMPQ3904)	$I_C = 100$ µA, $V_{CE} = 5.0$ V, $R_S = 1.0$kΩ, f=10 Hz to 15.7 kHz		5.0	dB

SWITCHING CHARACTERISTICS (except MMPQ3904)

Symbol	Parameter	Test Conditions	Min	Max	Units
t_d	Delay Time	$V_{CC} = 3.0$ V, $V_{BE} = 0.5$ V,		35	ns
t_r	Rise Time	$I_C = 10$ mA, $I_{B1} = 1.0$ mA		35	ns
t_s	Storage Time	$V_{CC} = 3.0$ V, $I_C = 10$mA		200	ns
t_f	Fall Time	$I_{B1} = I_{B2} = 1.0$ mA		50	ns

*Pulse Test: Pulse Width ≤ 300 µs, Duty Cycle ≤ 2.0%

Spice Model

NPN (Is=6.734f Xti=3 Eg=1.11 Vaf=74.03 Bf=416.4 Ne=1.259 Ise=6.734 Ikf=66.78m Xtb=1.5 Br=.7371 Nc=2
Isc=0 Ikr=0 Rc=1 Cjc=3.638p Mjc=.3085 Vjc=.75 Fc=.5 Cje=4.493p Mje=.2593 Vje=.75 Tr=239.5n Tf=301.2p
Itf=.4 Vtf=4 Xtf=2 Rb=10)

NPN General Purpose Amplifier

(continued)

Thermal Characteristics

TA = 25°C unless otherwise noted

Symbol	Characteristic	Max		Units
		2N3904	*PZT3904	
P$_D$	Total Device Dissipation	625	1,000	mW
	Derate above 25°C	5.0	8.0	mW/°C
R$_{θJC}$	Thermal Resistance, Junction to Case	83.3		°C/W
R$_{θJA}$	Thermal Resistance, Junction to Ambient	200	125	°C/W

Symbol	Characteristic	Max		Units
		**MMBT3904	MMPQ3904	
P$_D$	Total Device Dissipation	350	1,000	mW
	Derate above 25°C	2.8	8.0	mW/°C
R$_{θJA}$	Thermal Resistance, Junction to Ambient	357		°C/W
	Effective 4 Die		125	°C/W
	Each Die		240	°C/W

*Device mounted on FR-4 PCB 36 mm X 18 mm X 1.5 mm; mounting pad for the collector lead min. 6 cm^2.

**Device mounted on FR-4 PCB 1.6" X 1.6" X 0.06."

Typical Characteristics

Typical Pulsed Current Gain vs Collector Current

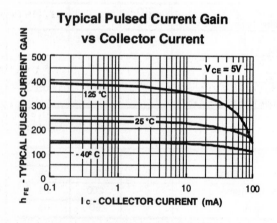

Collector-Emitter Saturation Voltage vs Collector Current

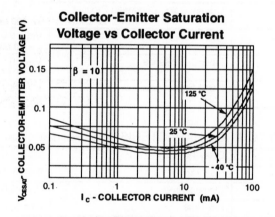

Base-Emitter Saturation Voltage vs Collector Current

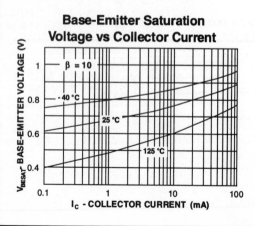

Base-Emitter ON Voltage vs Collector Current

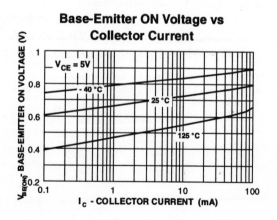

NPN General Purpose Amplifier
(continued)

Typical Characteristics (continued)

**Collector-Cutoff Current
vs Ambient Temperature**

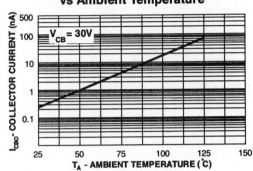

**Capacitance vs
Reverse Bias Voltage**

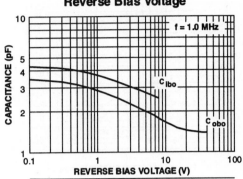

Noise Figure vs Frequency

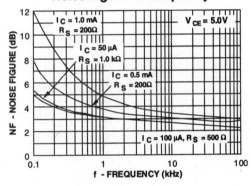

Noise Figure vs Source Resistance

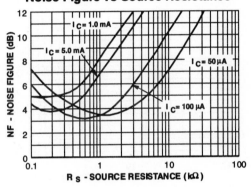

**Current Gain and Phase Angle
vs Frequency**

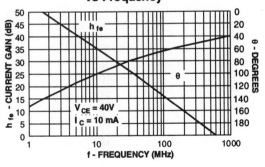

**Power Dissipation vs
Ambient Temperature**

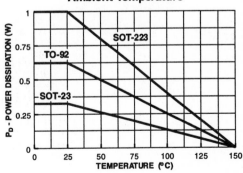

Typical Characteristics (continued)

Turn-On Time vs Collector Current

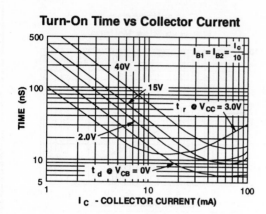

Rise Time vs Collector Current

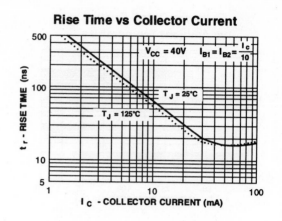

Storage Time vs Collector Current

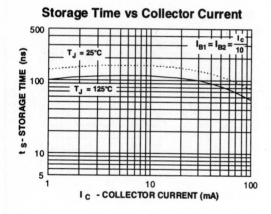

Fall Time vs Collector Current

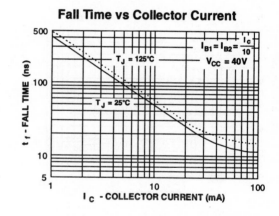

308

Test Circuits

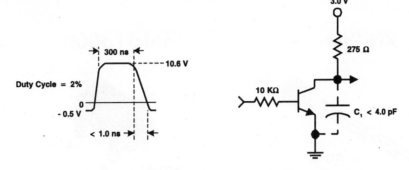

FIGURE 1: Delay and Rise Time Equivalent Test Circuit

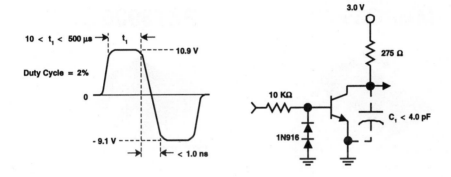

FIGURE 2: Storage and Fall Time Equivalent Test Circuit

FAIRCHILD

SEMICONDUCTOR ™

Discrete POWER & Signal
Technologies

2N3906

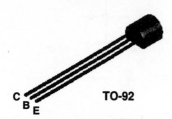

TO-92

MMBT3906

SOT-23
Mark: 2A

MMPQ3906

SOIC-16

PZT3906

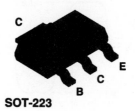

SOT-223

PNP General Purpose Amplifier

This device is designed for general purpose amplifier and switching applications at collector currents of 10 μA to 100 mA. Sourced from Process 66.

Absolute Maximum Ratings*   TA = 25°C unless otherwise noted

Symbol	Parameter	Value	Units
V_{CEO}	Collector-Emitter Voltage	40	V
V_{CBO}	Collector-Base Voltage	40	V
V_{EBO}	Emitter-Base Voltage	5.0	V
I_C	Collector Current - Continuous	200	mA
T_J, T_{stg}	Operating and Storage Junction Temperature Range	-55 to +150	°C

*These ratings are limiting values above which the serviceability of any semiconductor device may be impaired.

<u>NOTES</u>:
1) These ratings are based on a maximum junction temperature of 150 degrees C.
2) These are steady state limits. The factory should be consulted on applications involving pulsed or low duty cycle operations.

Electrical Characteristics TA = 25°C unless otherwise noted

Symbol	Parameter	Test Conditions	Min	Max	Units
OFF CHARACTERISTICS					
$V_{(BR)CEO}$	Collector-Emitter Breakdown Voltage*	$I_C = 1.0$ mA, $I_B = 0$	40		V
$V_{(BR)CBO}$	Collector-Base Breakdown Voltage	$I_C = 10$ µA, $I_E = 0$	40		V
$V_{(BR)EBO}$	Emitter-Base Breakdown Voltage	$I_E = 10$ µA, $I_C = 0$	5.0		V
I_{BL}	Base Cutoff Current	$V_{CE} = 30$ V, $V_{BE} = 3.0$ V		50	nA
I_{CEX}	Collector Cutoff Current	$V_{CE} = 30$ V, $V_{BE} = 3.0$ V		50	nA
ON CHARACTERISTICS					
h_{FE}	DC Current Gain *	$I_C = 0.1$ mA, $V_{CE} = 1.0$ V	60		
		$I_C = 1.0$ mA, $V_{CE} = 1.0$ V	80		
		$I_C = 10$ mA, $V_{CE} = 1.0$ V	100	300	
		$I_C = 50$ mA, $V_{CE} = 1.0$ V	60		
		$I_C = 100$ mA, $V_{CE} = 1.0$ V	30		
$V_{CE(sat)}$	Collector-Emitter Saturation Voltage	$I_C = 10$ mA, $I_B = 1.0$ mA		0.25	V
		$I_C = 50$ mA, $I_B = 5.0$ mA		0.4	V
$V_{BE(sat)}$	Base-Emitter Saturation Voltage	$I_C = 10$ mA, $I_B = 1.0$ mA	0.65	0.85	V
		$I_C = 50$ mA, $I_B = 5.0$ mA		0.95	V
SMALL SIGNAL CHARACTERISTICS					
f_T	Current Gain - Bandwidth Product	$I_C = 10$ mA, $V_{CE} = 20$ V, $f = 100$ MHz	250		MHz
C_{obo}	Output Capacitance	$V_{CB} = 5.0$ V, $I_E = 0$, $f = 100$ kHz		4.5	pF
C_{ibo}	Input Capacitance	$V_{EB} = 0.5$ V, $I_C = 0$, $f = 100$ kHz		10.0	pF
NF	Noise Figure (except MMPQ3906)	$I_C = 100$ µA, $V_{CE} = 5.0$ V, $R_S = 1.0$kΩ, $f = 10$ Hz to 15.7 kHz		4.0	dB
SWITCHING CHARACTERISTICS (except MMPQ3906)					
t_d	Delay Time	$V_{CC} = 3.0$ V, $V_{BE} = 0.5$ V,		35	ns
t_r	Rise Time	$I_C = 10$ mA, $I_{B1} = 1.0$ mA		35	ns
t_s	Storage Time	$V_{CC} = 3.0$ V, $I_C = 10$mA		225	ns
t_f	Fall Time	$I_{B1} = I_{B2} = 1.0$ mA		75	ns

*Pulse Test: Pulse Width ≤ 300 µs, Duty Cycle ≤ 2.0%

Spice Model

PNP (Is=1.41f Xti=3 Eg=1.11 Vaf=18.7 Bf=180.7 Ne=1.5 Ise=0 Ikf=80m Xtb=1.5 Br=4.977 Nc=2 Isc=0 Ikr=0
Rc=2.5 Cjc=9.728p Mjc=.5776 Vjc=.75 Fc=.5 Cje=8.063p Mje=.3677 Vje=.75 Tr=33.42n Tf=179.3p Itf=.4
Vtf=4 Xtf=6 Rb=10)

PNP General Purpose Amplifier

(continued)

Thermal Characteristics   TA = 25°C unless otherwise noted

Symbol	Characteristic	Max		Units
		2N3906	*PZT3906	
P$_D$	Total Device Dissipation	625	1,000	mW
	Derate above 25°C	5.0	8.0	mW/°C
R$_{\theta JC}$	Thermal Resistance, Junction to Case	83.3		°C/W
R$_{\theta JA}$	Thermal Resistance, Junction to Ambient	200	125	°C/W

Symbol	Characteristic	Max		Units
		**MMBT3906	MMPQ3906	
P$_D$	Total Device Dissipation	350	1,000	mW
	Derate above 25°C	2.8	8.0	mW/°C
R$_{\theta JA}$	Thermal Resistance, Junction to Ambient	357		°C/W
	Effective 4 Die		125	°C/W
	Each Die		240	°C/W

*Device mounted on FR-4 PCB 36 mm X 18 mm X 1.5 mm; mounting pad for the collector lead min. 6 cm².

**Device mounted on FR-4 PCB 1.6" X 1.6" X 0.06."

Typical Characteristics

Typical Pulsed Current Gain vs Collector Current

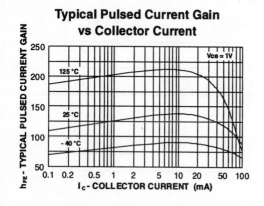

Collector-Emitter Saturation Voltage vs Collector Current

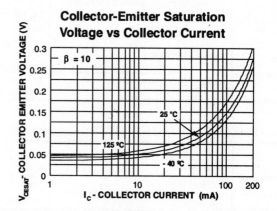

Base-Emitter Saturation Voltage vs Collector Current

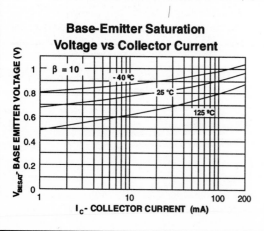

Base Emitter ON Voltage vs Collector Current

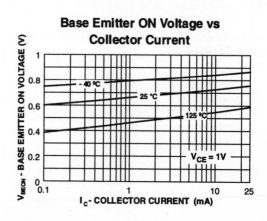

312

2N3906 / MMBT3906 / MMPQ3906 / PZT3906

PNP General Purpose Amplifier
(continued)

Typical Characteristics (continued)

Collector-Cutoff Current vs. Ambient Temperature

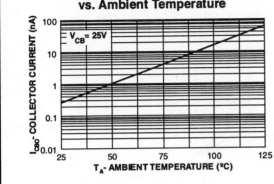

Common-Base Open Circuit Input and Output Capacitance vs Reverse Bias Voltage

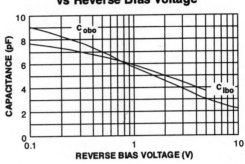

Noise Figure vs Frequency

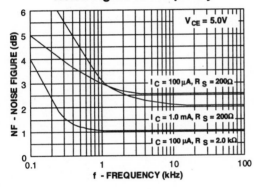

Noise Figure vs Source Resistance

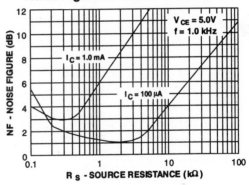

Switching Times vs Collector Current

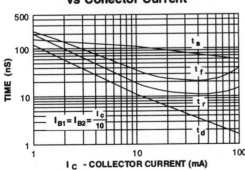

Turn On and Turn Off Times vs Collector Current

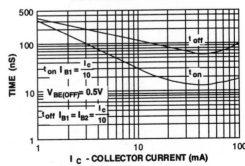

313

Typical Characteristics (continued)

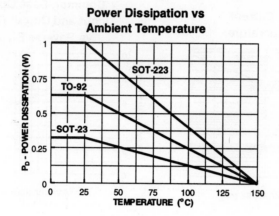

**Power Dissipation vs
Ambient Temperature**

JFET VHF Amplifier
N–Channel — Depletion

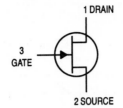

MPF102

CASE 29–04, STYLE 5
TO–92 (TO–226AA)

MAXIMUM RATINGS

Rating	Symbol	Value	Unit
Drain–Source Voltage	V_{DS}	25	Vdc
Drain–Gate Voltage	V_{DG}	25	Vdc
Gate–Source Voltage	V_{GS}	−25	Vdc
Gate Current	I_G	10	mAdc
Total Device Dissipation @ T_A = 25°C Derate above 25°C	P_D	350 2.8	mW mW/°C
Junction Temperature Range	T_J	125	°C
Storage Temperature Range	T_{stg}	−65 to +150	°C

ELECTRICAL CHARACTERISTICS (T_A = 25°C unless otherwise noted)

Characteristic	Symbol	Min	Max	Unit
OFF CHARACTERISTICS				
Gate–Source Breakdown Voltage (I_G = −10 µAdc, V_{DS} = 0)	$V_{(BR)GSS}$	−25	—	Vdc
Gate Reverse Current (V_{GS} = −15 Vdc, V_{DS} = 0) (V_{GS} = −15 Vdc, V_{DS} = 0, T_A = 100°C)	I_{GSS}	— —	−2.0 −2.0	nAdc µAdc
Gate–Source Cutoff Voltage (V_{DS} = 15 Vdc, I_D = 2.0 nAdc)	$V_{GS(off)}$	—	−8.0	Vdc
Gate–Source Voltage (V_{DS} = 15 Vdc, I_D = 0.2 mAdc)	V_{GS}	−0.5	−7.5	Vdc
ON CHARACTERISTICS				
Zero–Gate–Voltage Drain Current[1] (V_{DS} = 15 Vdc, V_{GS} = 0 Vdc)	I_{DSS}	2.0	20	mAdc
SMALL–SIGNAL CHARACTERISTICS				
Forward Transfer Admittance[1] (V_{DS} = 15 Vdc, V_{GS} = 0, f = 1.0 kHz) (V_{DS} = 15 Vdc, V_{GS} = 0, f = 100 MHz)	$|y_{fs}|$	2000 1600	7500 —	µmhos
Input Admittance (V_{DS} = 15 Vdc, V_{GS} = 0, f = 100 MHz)	$Re(y_{is})$	—	800	µmhos
Output Conductance (V_{DS} = 15 Vdc, V_{GS} = 0, f = 100 MHz)	$Re(y_{os})$	—	200	µmhos
Input Capacitance (V_{DS} = 15 Vdc, V_{GS} = 0, f = 1.0 MHz)	C_{iss}	—	7.0	pF
Reverse Transfer Capacitance (V_{DS} = 15 Vdc, V_{GS} = 0, f = 1.0 MHz)	C_{rss}	—	3.0	pF

1. Pulse Test; Pulse Width ≤ 630 ms, Duty Cycle ≤ 10%.

MOTOROLA

POWER GAIN

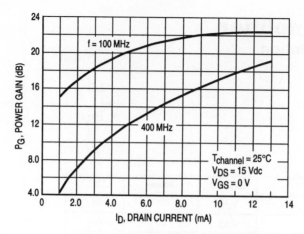

Figure 1. Effects of Drain Current

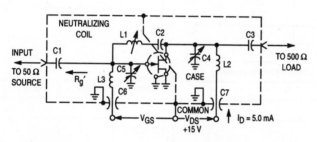

Reference Designation	VALUE	
	100 MHz	400 MHz
C1	7.0 pF	1.8 pF
C2	1000 pF	17 pF
C3	3.0 pF	1.0 pF
C4	1–12 pF	0.8–8.0 pF
C5	1–12 pF	0.8–8.0 pF
C6	0.0015 μF	0.001 μF
C7	0.0015 μF	0.001 μF
L1	3.0 μH*	0.2 μH**
L2	0.15 μH*	0.03 μH**
L3	0.14 μH*	0.022 μH**

Adjust V_{GS} for
$I_D = 50$ mA
$V_{GS} < 0$ Volts

NOTE: The noise source is a hot–cold body (AIL type 70 or equivalent) with a test receiver (AIL type 136 or equivalent).

*L1   17 turns, (approx. — depends upon circuit layout) AWG #28 enameled copper wire, close wound on 9/32″ ceramic coil form. Tuning provided by a powdered iron slug.
L2   4–1/2 turns, AWG #18 enameled copper wire, 5/16″ long, 3/8″ I.D. (AIR CORE).
L3   3–1/2 turns, AWG #18 enameled copper wire, 1/4″ long, 3/8″ I.D. (AIR CORE).

**L1   6 turns, (approx. — depends upon circuit layout) AWG #24 enameled copper wire, close wound on 7/32″ ceramic coil form. Tuning provided by an aluminum slug.
L2   1 turn, AWG #16 enameled copper wire, 3/8″ I.D. (AIR CORE).
L3   1/2 turn, AWG #16 enameled copper wire, 1/4″ I.D. (AIR CORE).

Figure 2. 100 MHz and 400 MHz Neutralized Test Circuit

NOISE FIGURE
($T_{channel} = 25°C$)

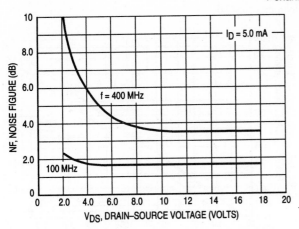

Figure 3. Effects of Drain–Source Voltage

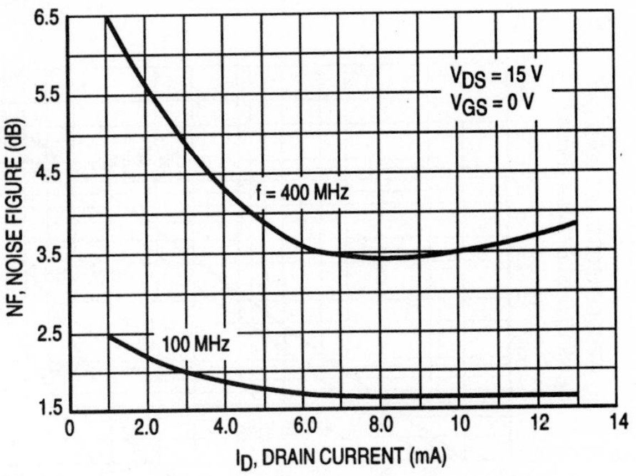

Figure 4. Effects of Drain Current

INTERMODULATION CHARACTERISTICS

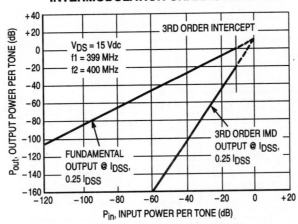

Figure 5. Third Order Intermodulation Distortion

Motorola Small–Signal Transistors, FETs and Diodes Device Data

COMMON SOURCE CHARACTERISTICS
ADMITTANCE PARAMETERS
(V_{DS} = 15 Vdc, $T_{channel}$ = 25°C)

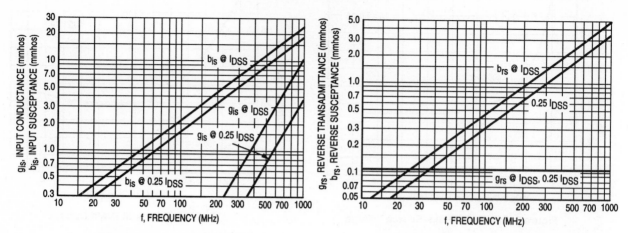

Figure 6. Input Admittance (y_{is})

Figure 7. Reverse Transfer Admittance (y_{rs})

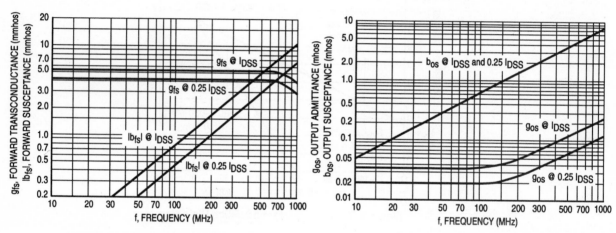

Figure 8. Forward Transadmittance (y_{fs})

Figure 9. Output Admittance (y_{os})

Motorola Small–Signal Transistors, FETs and Diodes Device Data

318

N *National Semiconductor*

November 1994

LM118/LM218/LM318
Operational Amplifiers

General Description

The LM118 series are precision high speed operational amplifiers designed for applications requiring wide bandwidth and high slew rate. They feature a factor of ten increase in speed over general purpose devices without sacrificing DC performance.

The LM118 series has internal unity gain frequency compensation. This considerably simplifies its application since no external components are necessary for operation. However, unlike most internally compensated amplifiers, external frequency compensation may be added for optimum performance. For inverting applications, feedforward compensation will boost the slew rate to over 150V/μs and almost double the bandwidth. Overcompensation can be used with the amplifier for greater stability when maximum bandwidth is not needed. Further, a single capacitor can be added to reduce the 0.1% settling time to under 1 μs.

The high speed and fast settling time of these op amps make them useful in A/D converters, oscillators, active filters, sample and hold circuits, or general purpose amplifiers. These devices are easy to apply and offer an order of magnitude better AC performance than industry standards such as the LM709.

The LM218 is identical to the LM118 except that the LM218 has its performance specified over a −25°C to +85°C temperature range. The LM318 is specified from 0°C to +70°C.

Features

- 15 MHz small signal bandwidth
- Guaranteed 50V/μs slew rate
- Maximum bias current of 250 nA
- Operates from supplies of ±5V to ±20V
- Internal frequency compensation
- Input and output overload protected
- Pin compatible with general purpose op amps

Connection Diagrams

Dual-In-Line Package

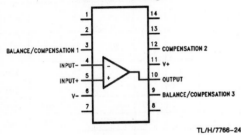

TL/H/7766–24

Top View

Order Number LM118J/883*
See NS Package Number J14A

Dual-In-Line Package

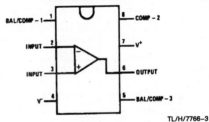

TL/H/7766–3

Top View

Order Number LM118J-8/883*,
LM318M or LM318N
See NS Package Number J08A, M08A or N08B

Metal Can Package**

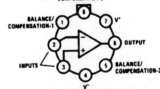

TL/H/7766–2

Top View

****Pin connections shown on schematic diagram
and typical applications are for TO-5 package.

Order Number LM118H, LM118H/883*,
LM218H or LM318H
See NS Package Number H08C

*Available per JM38510/10107.

©1995 National Semiconductor Corporation   TL/H/7766   RRD-B30M115/Printed in U.S.A.

Absolute Maximum Ratings

If Military/Aerospace specified devices are required, please contact the National Semiconductor Sales Office/Distributors for availability and specifications. (Note 6)

Supply Voltage	± 20V
Power Dissipation (Note 1)	500 mW
Differential Input Current (Note 2)	± 10 mA
Input Voltage (Note 3)	± 15V
Output Short-Circuit Duration	Continuous

Operating Temperature Range

LM118	$-55°$C to $+125°$C
LM218	$-25°$C to $+85°$C
LM318	$0°$C to $+70°$C
Storage Temperature Range	$-65°$C to $+150°$C

Lead Temperature (Soldering, 10 sec.)

Hermetic Package	300°C
Plastic Package	260°C

Soldering Information
Dual-In-Line Package

Soldering (10 sec.)	260°C

Small Outline Package

Vapor Phase (60 sec.)	215°C
Infrared (15 sec.)	220°C

See AN-450 "Surface Mounting Methods and Their Effect on Product Reliability" for other methods of soldering surface mount devices.

ESD Tolerance (Note 7)	2000V

Electrical Characteristics (Note 4)

Parameter	Conditions	LM118/LM218			LM318			Units
		Min	Typ	Max	Min	Typ	Max	
Input Offset Voltage	$T_A = 25°$C		2	4		4	10	mV
Input Offset Current	$T_A = 25°$C		6	50		30	200	nA
Input Bias Current	$T_A = 25°$C		120	250		150	500	nA
Input Resistance	$T_A = 25°$C	1	3		0.5	3		MΩ
Supply Current	$T_A = 25°$C		5	8		5	10	mA
Large Signal Voltage Gain	$T_A = 25°$C, $V_S = \pm 15$V $V_{OUT} = \pm 10$V, $R_L \geq 2$ kΩ	50	200		25	200		V/mV
Slew Rate	$T_A = 25°$C, $V_S = \pm 15$V, $A_V = 1$ (Note 5)	50	70		50	70		V/μs
Small Signal Bandwidth	$T_A = 25°$C, $V_S = \pm 15$V		15			15		MHz
Input Offset Voltage				6			15	mV
Input Offset Current				100			300	nA
Input Bias Current				500			750	nA
Supply Current	$T_A = 125°$C		4.5	7				mA
Large Signal Voltage Gain	$V_S = \pm 15$V, $V_{OUT} = \pm 10$V $R_L \geq 2$ kΩ	25			20			V/mV
Output Voltage Swing	$V_S = \pm 15$V, $R_L = 2$ kΩ	± 12	± 13		± 12	± 13		V
Input Voltage Range	$V_S = \pm 15$V	± 11.5			± 11.5			V
Common-Mode Rejection Ratio		80	100		70	100		dB
Supply Voltage Rejection Ratio		70	80		65	80		dB

Note 1: The maximum junction temperature of the LM118 is 150°C, the LM218 is 110°C, and the LM318 is 110°C. For operating at elevated temperatures, devices in the H08 package must be derated based on a thermal resistance of 160°C/W, junction to ambient, or 20°C/W, junction to case. The thermal resistance of the dual-in-line package is 100°C/W, junction to ambient.

Note 2: The inputs are shunted with back-to-back diodes for overvoltage protection. Therefore, excessive current will flow if a differential input voltage in excess of 1V is applied between the inputs unless some limiting resistance is used.

Note 3: For supply voltages less than ± 15V, the absolute maximum input voltage is equal to the supply voltage.

Note 4: These specifications apply for ± 5V $\leq V_S \leq \pm 20$V and $-55°$C $\leq T_A \leq +125°$C (LM118), $-25°$C $\leq T_A \leq +85°$C (LM218), and $0°$C $\leq T_A \leq +70°$C (LM318). Also, power supplies must be bypassed with 0.1 μF disc capacitors.

Note 5: Slew rate is tested with $V_S = \pm 15$V. The LM118 is in a unity-gain non-inverting configuration. V_{IN} is stepped from -7.5V to $+7.5$V and vice versa. The slew rates between -5.0V and $+5.0$V and vice versa are tested and guaranteed to exceed 50V/μs.

Note 6: Refer to RETS118X for LM118H and LM118J military specifications.

Note 7: Human body model, 1.5 kΩ in series with 100 pF.

320

Typical Performance Characteristics LM118, LM218

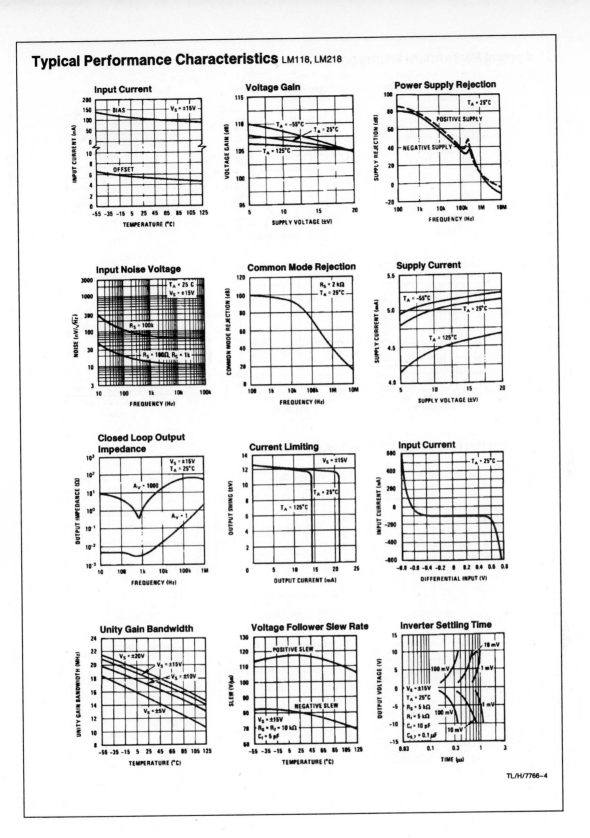

TL/H/7766–4

Typical Performance Characteristics LM118, LM218 (Continued)

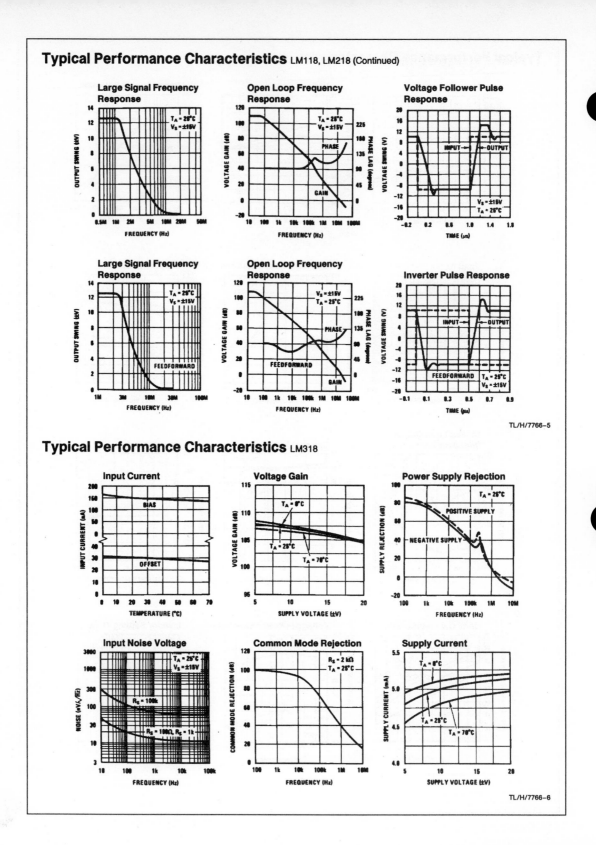

TL/H/7766–5

Typical Performance Characteristics LM318

TL/H/7766–6

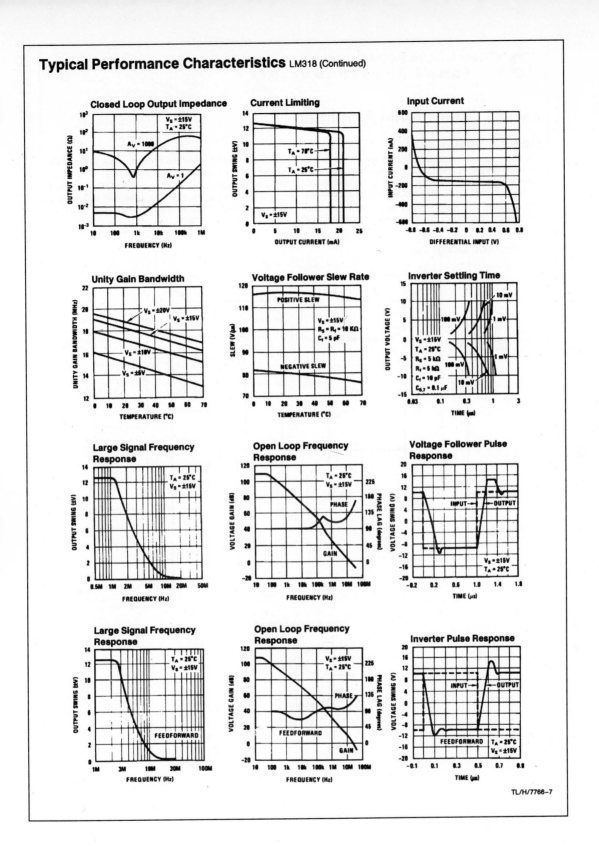

TL/H/7766–7

National Semiconductor

November 1994

LM741 Operational Amplifier

General Description

The LM741 series are general purpose operational amplifiers which feature improved performance over industry standards like the LM709. They are direct, plug-in replacements for the 709C, LM201, MC1439 and 748 in most applications.

The amplifiers offer many features which make their application nearly foolproof: overload protection on the input and output, no latch-up when the common mode range is exceeded, as well as freedom from oscillations.

The LM741C/LM741E are identical to the LM741/LM741A except that the LM741C/LM741E have their performance guaranteed over a 0°C to +70°C temperature range, instead of −55°C to +125°C.

Schematic Diagram

TL/H/9341–1

Offset Nulling Circuit

TL/H/9341–7

RRD-B30M115/Printed in U. S. A.

Absolute Maximum Ratings

If Military/Aerospace specified devices are required, please contact the National Semiconductor Sales Office/Distributors for availability and specifications.

(Note 5)

	LM741A	LM741E	LM741	LM741C
Supply Voltage	±22V	±22V	±22V	±18V
Power Dissipation (Note 1)	500 mW	500 mW	500 mW	500 mW
Differential Input Voltage	±30V	±30V	±30V	±30V
Input Voltage (Note 2)	±15V	±15V	±15V	±15V
Output Short Circuit Duration	Continuous	Continuous	Continuous	Continuous
Operating Temperature Range	−55°C to +125°C	0°C to +70°C	−55°C to +125°C	0°C to +70°C
Storage Temperature Range	−65°C to +150°C	−65°C to +150°C	−65°C to +150°C	−65°C to +150°C
Junction Temperature	150°C	100°C	150°C	100°C
Soldering Information				
N-Package (10 seconds)	260°C	260°C	260°C	260°C
J- or H-Package (10 seconds)	300°C	300°C	300°C	300°C
M-Package				
Vapor Phase (60 seconds)	215°C	215°C	215°C	215°C
Infrared (15 seconds)	215°C	215°C	215°C	215°C

See AN-450 "Surface Mounting Methods and Their Effect on Product Reliability" for other methods of soldering surface mount devices.

| ESD Tolerance (Note 6) | 400V | 400V | 400V | 400V |

Electrical Characteristics (Note 3)

Parameter	Conditions	LM741A/LM741E			LM741			LM741C			Units
		Min	Typ	Max	Min	Typ	Max	Min	Typ	Max	
Input Offset Voltage	$T_A = 25°C$ $R_S \leq 10\ k\Omega$ $R_S \leq 50\Omega$		0.8	3.0	1.0		5.0	2.0		6.0	mV mV
	$T_{AMIN} \leq T_A \leq T_{AMAX}$ $R_S \leq 50\Omega$ $R_S \leq 10\ k\Omega$			4.0			6.0			7.5	mV mV
Average Input Offset Voltage Drift				15							μV/°C
Input Offset Voltage Adjustment Range	$T_A = 25°C$, $V_S = ±20V$	±10				±15			±15		mV
Input Offset Current	$T_A = 25°C$		3.0	30		20	200		20	200	nA
	$T_{AMIN} \leq T_A \leq T_{AMAX}$			70		85	500			300	nA
Average Input Offset Current Drift				0.5							nA/°C
Input Bias Current	$T_A = 25°C$		30	80		80	500		80	500	nA
	$T_{AMIN} \leq T_A \leq T_{AMAX}$			0.210			1.5			0.8	μA
Input Resistance	$T_A = 25°C$, $V_S = ±20V$	1.0	6.0		0.3	2.0		0.3	2.0		MΩ
	$T_{AMIN} \leq T_A \leq T_{AMAX}$, $V_S = ±20V$	0.5									MΩ
Input Voltage Range	$T_A = 25°C$							±12	±13		V
	$T_{AMIN} \leq T_A \leq T_{AMAX}$				±12	±13					V
Large Signal Voltage Gain	$T_A = 25°C$, $R_L \geq 2\ k\Omega$ $V_S = ±20V$, $V_O = ±15V$ $V_S = ±15V$, $V_O = ±10V$	50			50	200		20	200		V/mV V/mV
	$T_{AMIN} \leq T_A \leq T_{AMAX}$, $R_L \geq 2\ k\Omega$, $V_S = ±20V$, $V_O = ±15V$ $V_S = ±15V$, $V_O = ±10V$ $V_S = ±5V$, $V_O = ±2V$	32 10			25			15			V/mV V/mV V/mV

Electrical Characteristics (Note 3) (Continued)

Parameter	Conditions	LM741A/LM741E Min	LM741A/LM741E Typ	LM741A/LM741E Max	LM741 Min	LM741 Typ	LM741 Max	LM741C Min	LM741C Typ	LM741C Max	Units
Output Voltage Swing	$V_S = \pm 20V$ $R_L \geq 10\ k\Omega$ $R_L \geq 2\ k\Omega$	± 16 ± 15									V V
	$V_S = \pm 15V$ $R_L \geq 10\ k\Omega$ $R_L \geq 2\ k\Omega$				± 12 ± 10	± 14 ± 13		± 12 ± 10	± 14 ± 13		V V
Output Short Circuit Current	$T_A = 25°C$ $T_{AMIN} \leq T_A \leq T_{AMAX}$	10 10	25	35 40	25			25			mA mA
Common-Mode Rejection Ratio	$T_{AMIN} \leq T_A \leq T_{AMAX}$ $R_S \leq 10\ k\Omega$, $V_{CM} = \pm 12V$ $R_S \leq 50\Omega$, $V_{CM} = \pm 12V$	80	95		70	90		70	90		dB dB
Supply Voltage Rejection Ratio	$T_{AMIN} \leq T_A \leq T_{AMAX}$, $V_S = \pm 20V$ to $V_S = \pm 5V$ $R_S \leq 50\Omega$ $R_S \leq 10\ k\Omega$	86	96		77	96		77	96		dB dB
Transient Response Rise Time Overshoot	$T_A = 25°C$, Unity Gain		0.25 6.0	0.8 20		0.3 5			0.3 5		µs %
Bandwidth (Note 4)	$T_A = 25°C$	0.437	1.5								MHz
Slew Rate	$T_A = 25°C$, Unity Gain	0.3	0.7			0.5			0.5		V/µs
Supply Current	$T_A = 25°C$					1.7	2.8		1.7	2.8	mA
Power Consumption	$T_A = 25°C$ $V_S = \pm 20V$ $V_S = \pm 15V$		80	150		50	85		50	85	mW mW
LM741A	$V_S = \pm 20V$ $T_A = T_{AMIN}$ $T_A = T_{AMAX}$			165 135							mW mW
LM741E	$V_S = \pm 20V$ $T_A = T_{AMIN}$ $T_A = T_{AMAX}$			150 150							mW mW
LM741	$V_S = \pm 15V$ $T_A = T_{AMIN}$ $T_A = T_{AMAX}$					60 45	100 75				mW mW

Note 1: For operation at elevated temperatures, these devices must be derated based on thermal resistance, and T_j max. (listed under "Absolute Maximum Ratings"). $T_j = T_A + (\theta_{jA} P_D)$.

Thermal Resistance	Cerdip (J)	DIP (N)	HO8 (H)	SO-8 (M)
θ_{jA} (Junction to Ambient)	100°C/W	100°C/W	170°C/W	195°C/W
θ_{jC} (Junction to Case)	N/A	N/A	25°C/W	N/A

Note 2: For supply voltages less than $\pm 15V$, the absolute maximum input voltage is equal to the supply voltage.

Note 3: Unless otherwise specified, these specifications apply for $V_S = \pm 15V$, $-55°C \leq T_A \leq +125°C$ (LM741/LM741A). For the LM741C/LM741E, these specifications are limited to $0°C \leq T_A \leq +70°C$.

Note 4: Calculated value from: BW (MHz) = 0.35/Rise Time(µs).

Note 5: For military specifications see RETS741X for LM741 and RETS741AX for LM741A.

Note 6: Human body model, 1.5 kΩ in series with 100 pF.

Connection Diagrams

Metal Can Package

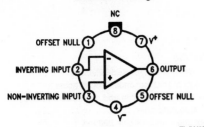

TL/H/9341–2

**Order Number LM741H, LM741H/883*,
LM741AH/883 or LM741CH
See NS Package Number H08C**

Ceramic Dual-In-Line Package

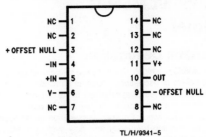

TL/H/9341–5

Order Number LM741J-14/883*, LM741AJ-14/883
See NS Package Number J14A**

*also available per JM38510/10101
**also available per JM38510/10102

Dual-In-Line or S.O. Package

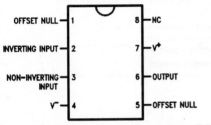

TL/H/9341–3

**Order Number LM741J, LM741J/883,
LM741CM, LM741CN or LM741EN
See NS Package Number J08A, M08A or N08E**

Ceramic Flatpak

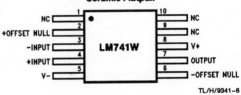

TL/H/9341–6

**Order Number LM741W/883
See NS Package Number W10A**

*LM741H is available per JM38510/10101

National *Semiconductor*

May 1997

LM555/LM555C
Timer

General Description

The LM555 is a highly stable device for generating accurate time delays or oscillation. Additional terminals are provided for triggering or resetting if desired. In the time delay mode of operation, the time is precisely controlled by one external resistor and capacitor. For astable operation as an oscillator, the free running frequency and duty cycle are accurately controlled with two external resistors and one capacitor. The circuit may be triggered and reset on falling waveforms, and the output circuit can source or sink up to 200 mA or drive TTL circuits.

Features

- Direct replacement for SE555/NE555
- Timing from microseconds through hours
- Operates in both astable and monostable modes
- Adjustable duty cycle
- Output can source or sink 200 mA
- Output and supply TTL compatible
- Temperature stability better than 0.005% per °C
- Normally on and normally off output
- Available in 8 pin MSOP package

Applications

- Precision timing
- Pulse generation
- Sequential timing
- Time delay generation
- Pulse width modulation
- Pulse position modulation
- Linear ramp generator

Schematic Diagram

DS007851-1

www.national.com

Absolute Maximum Ratings (Note 2)

If Military/Aerospace specified devices are required, please contact the National Semiconductor Sales Office/ Distributors for availability and specifications.

Supply Voltage	+18V
Power Dissipation (Note 3)	
LM555H, LM555CH	760 mW
LM555, LM555CN	1180 mW
LM555CMM	613 mW
Operating Temperature Ranges	
LM555C	0°C to +70°C
LM555	−55°C to +125°C

Storage Temperature Range	−65°C to +150°C
Soldering Information	
Dual-In-Line Package	
Soldering (10 Seconds)	260°C
Small Outline Packages	
(SOIC and MSOP)	
Vapor Phase (60 Seconds)	215°C
Infrared (15 Seconds)	220°C

See AN-450 "Surface Mounting Methods and Their Effect on Product Reliability" for other methods of soldering surface mount devices.

Electrical Characteristics (Notes 1, 2)

(T_A = 25°C, V_{CC} = +5V to +15V, unless othewise specified)

Parameter	Conditions	Limits						Units
		LM555			LM555C			
		Min	Typ	Max	Min	Typ	Max	
Supply Voltage		4.5		18	4.5		16	V
Supply Current	V_{CC} = 5V, R_L = ∞		3	5		3	6	mA
	V_{CC} = 15V, R_L = ∞		10	12		10	15	mA
	(Low State) (Note 4)							
Timing Error, Monostable								
Initial Accuracy			0.5			1		%
Drift with Temperature	R_A = 1k to 100 kΩ,		30			50		ppm/°C
	C = 0.1 μF, (Note 5)							
Accuracy over Temperature			1.5			1.5		%
Drift with Supply			0.05			0.1		%/V
Timing Error, Astable								
Initial Accuracy			1.5			2.25		%
Drift with Temperature	R_A, R_B = 1k to 100 kΩ,		90			150		ppm/°C
	C = 0.1 μF, (Note 5)							
Accuracy over Temperature			2.5			3.0		%
Drift with Supply			0.15			0.30		%/V
Threshold Voltage			0.667			0.667		x V_{CC}
Trigger Voltage	V_{CC} = 15V	4.8	5	5.2		5		V
	V_{CC} = 5V	1.45	1.67	1.9		1.67		V
Trigger Current			0.01	0.5		0.5	0.9	μA
Reset Voltage		0.4	0.5	1	0.4	0.5	1	V
Reset Current			0.1	0.4		0.1	0.4	mA
Threshold Current	(Note 6)		0.1	0.25		0.1	0.25	μA
Control Voltage Level	V_{CC} = 15V	9.6	10	10.4	9	10	11	V
	V_{CC} = 5V	2.9	3.33	3.8	2.6	3.33	4	V
Pin 7 Leakage Output High			1	100		1	100	nA
Pin 7 Sat (Note 7)								
Output Low	V_{CC} = 15V, I_7 = 15 mA		150			180		mV
Output Low	V_{CC} = 4.5V, I_7 = 4.5 mA		70	100		80	200	mV

Electrical Characteristics (Notes 1, 2) (Continued)

(T_A = 25°C, V_{CC} = +5V to +15V, unless othewise specified)

Parameter	Conditions	LM555			LM555C			Units
		Min	Typ	Max	Min	Typ	Max	
Output Voltage Drop (Low)	V_{CC} = 15V							
	I_{SINK} = 10 mA		0.1	0.15		0.1	0.25	V
	I_{SINK} = 50 mA		0.4	0.5		0.4	0.75	V
	I_{SINK} = 100 mA		2	2.2		2	2.5	V
	I_{SINK} = 200 mA		2.5			2.5		V
	V_{CC} = 5V							
	I_{SINK} = 8 mA		0.1	0.25				V
	I_{SINK} = 5 mA					0.25	0.35	V
Output Voltage Drop (High)	I_{SOURCE} = 200 mA, V_{CC} = 15V		12.5			12.5		V
	I_{SOURCE} = 100 mA, V_{CC} = 15V	13	13.3		12.75	13.3		V
	V_{CC} = 5V	3	3.3		2.75	3.3		V
Rise Time of Output			100			100		ns
Fall Time of Output			100			100		ns

Note 1: All voltages are measured with respect to the ground pin, unless otherwise specified.

Note 2: Absolute Maximum Ratings indicate limits beyond which damage to the device may occur. Operating Ratings indicate conditions for which the device is functional, but do not guarantee specific performance limits. Electrical Characteristics state DC and AC electrical specifications under particular test conditions which guarantee specific performance limits. This assumes that the device is within the Operating Ratings. Specifications are not guaranteed for parameters where no limit is given, however, the typical value is a good indication of device performance.

Note 3: For operating at elevated temperatures the device must be derated above 25°C based on a +150°C maximum junction temperature and a thermal resistance of 164°C/W (T0-5), 106°C/W (DIP), 170°C/W (S0-8), and 204°C/W (MSOP) junction to ambient.

Note 4: Supply current when output high typically 1 mA less at V_{CC} = 5V.

Note 5: Tested at V_{CC} = 5V and V_{CC} = 15V.

Note 6: This will determine the maximum value of $R_A + R_B$ for 15V operation. The maximum total ($R_A + R_B$) is 20 MΩ.

Note 7: No protection against excessive pin 7 current is necessary providing the package dissipation rating will not be exceeded.

Note 8: Refer to RETS555X drawing of military LM555H and LM555J versions for specifications.

Connection Diagrams

Metal Can Package

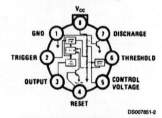

DS007851-2

Top View
Order Number LM555H or LM555CH
See NS Package Number H08C

**Dual-In-Line, Small Outline
and Molded Mini Small Outline Packages**

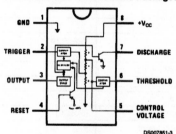

DS007851-3

Top View
**Order Number LM555J, LM555CJ,
LM555CM, LM555CMM or LM555CN**
**See NS Package Number J08A, M08A, MUA08A or
N08E**

Typical Performance Characteristics

Minimuim Pulse Width Required for Triggering

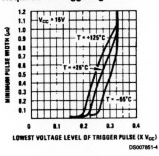

DS007851-4

Supply Current vs Supply Voltage

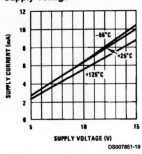

DS007851-19

High Output Voltage vs Output Source Current

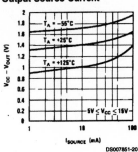

DS007851-20

Low Output Voltage vs Output Sink Current

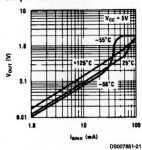

DS007851-21

Low Output Voltage vs Output Sink Current

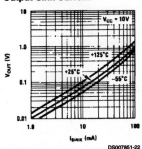

DS007851-22

Low Output Voltage vs Output Sink Current

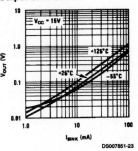

DS007851-23

Output Propagation Delay vs Voltage Level of Trigger Pulse

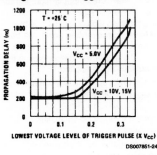

DS007851-24

Output Propagation Delay vs Voltage Level of Trigger Pulse

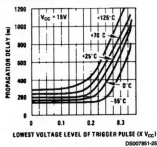

DS007851-25

Discharge Transistor (Pin 7) Voltage vs Sink Current

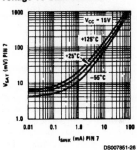

DS007851-26

331

Typical Performance Characteristics (Continued)

**Discharge Transistor (Pin 7)
Voltage vs Sink Current**

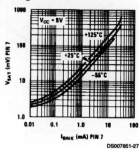

DS007851-27

Applications Information

MONOSTABLE OPERATION

In this mode of operation, the timer functions as a one-shot (*Figure 1*). The external capacitor is initially held discharged by a transistor inside the timer. Upon application of a negative trigger pulse of less than 1/3 V_{CC} to pin 2, the flip-flop is set which both releases the short circuit across the capacitor and drives the output high.

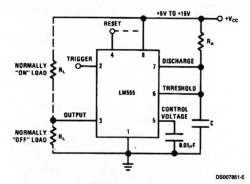

FIGURE 1. Monostable

The voltage across the capacitor then increases exponentially for a period of t = 1.1 R_A C, at the end of which time the voltage equals 2/3 V_{CC}. The comparator then resets the flip-flop which in turn discharges the capacitor and drives the output to its low state. *Figure 2* shows the waveforms generated in this mode of operation. Since the charge and the threshold level of the comparator are both directly proportional to supply voltage, the timing internal is independent of supply.

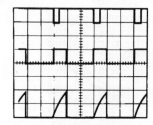

DS007851-6

V_{CC} = 5V   Top Trace: Input 5V/Div.
TIME = 0.1 ms/DIV.   Middle Trace: Output 5V/Div.
R_A = 9.1 kΩ   Bottom Trace: Capacitor Voltage 2V/Div.
C = 0.01 μF

FIGURE 2. Monostable Waveforms

During the timing cycle when the output is high, the further application of a trigger pulse will not effect the circuit so long as the trigger input is returned high at least 10 μs before the end of the timing interval. However the circuit can be reset during this time by the application of a negative pulse to the reset terminal (pin 4). The output will then remain in the low state until a trigger pulse is again applied.

When the reset function is not in use, it is recommended that it be connected to V_{CC} to avoid any possibility of false triggering.

Figure 3 is a nomograph for easy determination of R, C values for various time delays.

NOTE: In monostable operation, the trigger should be driven high before the end of timing cycle.

Applications Information (Continued)

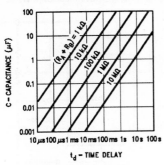

FIGURE 3. Time Delay

ASTABLE OPERATION

If the circuit is connected as shown in *Figure 4* (pins 2 and 6 connected) it will trigger itself and free run as a multivibrator. The external capacitor charges through $R_A + R_B$ and discharges through R_B. Thus the duty cycle may be precisely set by the ratio of these two resistors.

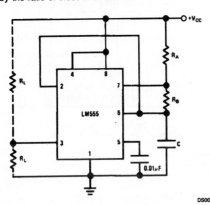

FIGURE 4. Astable

In this mode of operation, the capacitor charges and discharges between $1/3\ V_{CC}$ and $2/3\ V_{CC}$. As in the triggered mode, the charge and discharge times, and therefore the frequency are independent of the supply voltage.

Figure 5 shows the waveforms generated in this mode of operation.

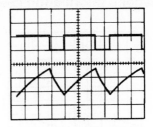

V_{CC} = 5V
TIME = 20 μs/DIV.
R_A = 3.9 kΩ
R_B = 3 kΩ
C = 0.01 μF

Top Trace: Output 5V/Div.
Bottom Trace: Capacitor Voltage 1V/Div.

FIGURE 5. Astable Waveforms

The charge time (output high) is given by:
$$t_1 = 0.693\ (R_A + R_B)\ C$$
And the discharge time (output low) by:
$$t_2 = 0.693\ (R_B)\ C$$
Thus the total period is:
$$T = t_1 + t_2 = 0.693\ (R_A + 2R_B)\ C$$
The frequency of oscillation is:

$$f = \frac{1}{T} = \frac{1.44}{(R_A + 2\ R_B)\ C}$$

Figure 6 may be used for quick determination of these RC values.

The duty cycle is:

$$D = \frac{R_B}{R_A + 2R_B}$$

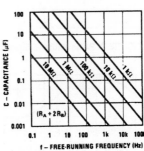

FIGURE 6. Free Running Frequency

FREQUENCY DIVIDER

The monostable circuit of *Figure 1* can be used as a frequency divider by adjusting the length of the timing cycle. *Figure 7* shows the waveforms generated in a divide by three circuit.

www.national.com

333

Applications Information (Continued)

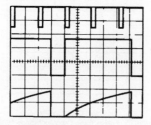

V_{CC} = 5V   Top Trace: Input 4V/Div.
TIME = 20 μs/DIV.   Middle Trace: Output 2V/Div.
R_A = 9.1 kΩ   Bottom Trace: Capacitor 2V/Div.
C = 0.01 μF

FIGURE 7. Frequency Divider

PULSE WIDTH MODULATOR

When the timer is connected in the monostable mode and triggered with a continuous pulse train, the output pulse width can be modulated by a signal applied to pin 5. *Figure 8* shows the circuit, and in *Figure 9* are some waveform examples.

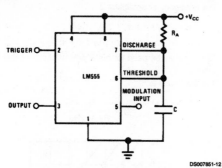

FIGURE 8. Pulse Width Modulator

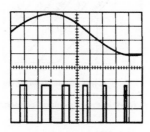

V_{CC} = 5V   Top Trace: Modulation 1V/Div.
TIME = 0.2 ms/DIV.   Bottom Trace: Output Voltage 2V/Div.
R_A = 9.1 kΩ
C = 0.01 μF

FIGURE 9. Pulse Width Modulator

PULSE POSITION MODULATOR

This application uses the timer connected for astable operation, as in *Figure 10*, with a modulating signal again applied to the control voltage terminal. The pulse position varies with the modulating signal, since the threshold voltage and hence the time delay is varied. *Figure 11* shows the waveforms generated for a triangle wave modulation signal.

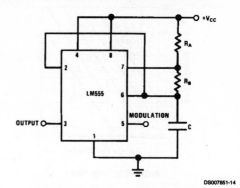

FIGURE 10. Pulse Position Modulator

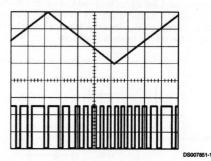

V_{CC} = 5V   Top Trace: Modulation Input 1V/Div.
TIME = 0.1 ms/DIV.   Bottom Trace: Output 2V/Div.
R_A = 3.9 kΩ
R_B = 3 kΩ
C = 0.01 μF

FIGURE 11. Pulse Position Modulator

LINEAR RAMP

When the pullup resistor, R_A, in the monostable circuit is replaced by a constant current source, a linear ramp is generated. *Figure 12* shows a circuit configuration that will perform this function.

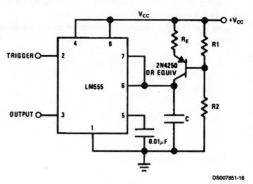

FIGURE 12.

Applications Information (Continued)

Figure 13 shows waveforms generated by the linear ramp. The time interval is given by:

$$T = \frac{2/3\ V_{CC}\ R_E\ (R_1 + R_2)\ C}{R_1\ V_{CC} - V_{BE}\ (R_1 + R_2)}$$

$$V_{BE} \cong 0.6V$$

$$V_{BE} \cong 0.6V$$

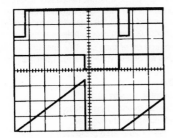

DS007851-17

V_{CC} = 5V   Top Trace: Input 3V/Div.
TIME = 20 μs/DIV.   Middle Trace: Output 5V/Div.
R_1 = 47 kΩ   Bottom Trace: Capacitor Voltage 1V/Div.
R_2 = 100 kΩ
R_E = 2.7 kΩ
C = 0.01 μF

FIGURE 13. Linear Ramp

50% DUTY CYCLE OSCILLATOR

For a 50% duty cycle, the resistors R_A and R_B may be connected as in *Figure 14*. The time period for the output high is the same as previous, $t_1 = 0.693\ R_A\ C$. For the output low it is t_2 =

$$\left[(R_A R_B)/(R_A + R_B) \right] C\ ln \left[\frac{R_B - 2R_A}{2R_B - R_A} \right]$$

Thus the frequency of oscillation is

$$f = \frac{1}{t_1 + t_2}$$

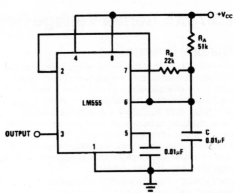

DS007851-18

FIGURE 14. 50% Duty Cycle Oscillator

Note that this circuit will not oscillate if R_B is greater than 1/2 R_A because the junction of R_A and R_B cannot bring pin 2 down to 1/3 V_{CC} and trigger the lower comparator.

ADDITIONAL INFORMATION

Adequate power supply bypassing is necessary to protect associated circuitry. Minimum recommended is 0.1 μF in parallel with 1 μF electrolytic.

Lower comparator storage time can be as long as 10 μs when pin 2 is driven fully to ground for triggering. This limits the monostable pulse width to 10 μs minimum.

Delay time reset to output is 0.47 μs typical. Minimum reset pulse width must be 0.3 μs, typical.

Pin 7 current switches within 30 ns of the output (pin 3) voltage.

National Semiconductor

February 1995

LM78XX Series Voltage Regulators

General Description

The LM78XX series of three terminal regulators is available with several fixed output voltages making them useful in a wide range of applications. One of these is local on card regulation, eliminating the distribution problems associated with single point regulation. The voltages available allow these regulators to be used in logic systems, instrumentation, HiFi, and other solid state electronic equipment. Although designed primarily as fixed voltage regulators these devices can be used with external components to obtain adjustable voltages and currents.

The LM78XX series is available in an aluminum TO-3 package which will allow over 1.0A load current if adequate heat sinking is provided. Current limiting is included to limit the peak output current to a safe value. Safe area protection for the output transistor is provided to limit internal power dissipation. If internal power dissipation becomes too high for the heat sinking provided, the thermal shutdown circuit takes over preventing the IC from overheating.

Considerable effort was expanded to make the LM78XX series of regulators easy to use and mininize the number of external components. It is not necessary to bypass the output, although this does improve transient response. Input bypassing is needed only if the regulator is located far from the filter capacitor of the power supply.

For output voltage other than 5V, 12V and 15V the LM117 series provides an output voltage range from 1.2V to 57V.

Features

- Output current in excess of 1A
- Internal thermal overload protection
- No external components required
- Output transistor safe area protection
- Internal short circuit current limit
- Available in the aluminum TO-3 package

Voltage Range

LM7805C	5V
LM7812C	12V
LM7815C	15V

Schematic and Connection Diagrams

TL/H/7746–1

Metal Can Package TO-3 (K) Aluminum

TL/H/7746–2

Bottom View

Order Number LM7805CK, LM7812CK or LM7815CK See NS Package Number KC02A

Plastic Package TO-220 (T)

TL/H/7746–3

Top View

Order Number LM7805CT, LM7812CT or LM7815CT See NS Package Number T03B

Absolute Maximum Ratings

Input Voltage (V_O = 5V, 12V and 15V)	35V
Internal Power Dissipation (Note 1)	Internally Limited
Operating Temperature Range (T_A)	0°C to +70°C
Maximum Junction Temperature	
(K Package)	150°C
(T Package)	150°C
Storage Temperature Range	−65°C to +150°C
Lead Temperature (Soldering, 10 sec.)	
TO-3 Package K	300°C
TO-220 Package T	230°C

Electrical Characteristics LM78XXC (Note 2) 0°C ≤ Tj ≤ 125°C unless otherwise noted.

	Output Voltage		5V			12V			15V			
	Input Voltage (unless otherwise noted)		10V			19V			23V			Units
Symbol	Parameter	Conditions	Min	Typ	Max	Min	Typ	Max	Min	Typ	Max	
V_O	Output Voltage	Tj = 25°C, 5 mA ≤ I_O ≤ 1A	4.8	5	5.2	11.5	12	12.5	14.4	15	15.6	V
		P_D ≤ 15W, 5 mA ≤ I_O ≤ 1A	4.75		5.25	11.4		12.6	14.25		15.75	V
		V_{MIN} ≤ V_{IN} ≤ V_{MAX}	(7.5 ≤ V_{IN} ≤ 20)			(14.5 ≤ V_{IN} ≤ 27)			(17.5 ≤ V_{IN} ≤ 30)			V
ΔV_O	Line Regulation	I_O = 500 mA, Tj = 25°C, ΔV_{IN}		3	50		4	120		4	150	mV
			(7 ≤ V_{IN} ≤ 25)			(14.5 ≤ V_{IN} ≤ 30)			(17.5 ≤ V_{IN} ≤ 30)			V
		0°C ≤ Tj ≤ +125°C, ΔV_{IN}			50			120			150	mV
			(8 ≤ V_{IN} ≤ 20)			(15 ≤ V_{IN} ≤ 27)			(18.5 ≤ V_{IN} ≤ 30)			V
		I_O ≤ 1A, Tj = 25°C, ΔV_{IN}			50			120			150	mV
			(7.5 ≤ V_{IN} ≤ 20)			(14.6 ≤ V_{IN} ≤ 27)			(17.7 ≤ V_{IN} ≤ 30)			V
		0°C ≤ Tj ≤ +125°C, ΔV_{IN}			25			60			75	mV
			(8 ≤ V_{IN} ≤ 12)			(16 ≤ V_{IN} ≤ 22)			(20 ≤ V_{IN} ≤ 26)			V
ΔV_O	Load Regulation	Tj = 25°C, 5 mA ≤ I_O ≤ 1.5A		10	50		12	120		12	150	mV
		250 mA ≤ I_O ≤ 750 mA			25			60			75	mV
		5 mA ≤ I_O ≤ 1A, 0°C ≤ Tj ≤ +125°C			50			120			150	mV
I_Q	Quiescent Current	I_O ≤ 1A, Tj = 25°C			8			8			8	mA
		0°C ≤ Tj ≤ +125°C			8.5			8.5			8.5	mA
ΔI_Q	Quiescent Current Change	5 mA ≤ I_O ≤ 1A			0.5			0.5			0.5	mA
		Tj = 25°C, I_O ≤ 1A, V_{MIN} ≤ V_{IN} ≤ V_{MAX}			1.0			1.0			1.0	mA
			(7.5 ≤ V_{IN} ≤ 20)			(14.8 ≤ V_{IN} ≤ 27)			(17.9 ≤ V_{IN} ≤ 30)			V
		I_O ≤ 500 mA, 0°C ≤ Tj ≤ +125°C, V_{MIN} ≤ V_{IN} ≤ V_{MAX}			1.0			1.0			1.0	mA
			(7 ≤ V_{IN} ≤ 25)			(14.5 ≤ V_{IN} ≤ 30)			(17.5 ≤ V_{IN} ≤ 30)			V
V_N	Output Noise Voltage	T_A = 25°C, 10 Hz ≤ f ≤ 100 kHz		40			75			90		µV
$\dfrac{\Delta V_{IN}}{\Delta V_{OUT}}$	Ripple Rejection	I_O ≤ 1A, Tj = 25°C or f = 120 Hz I_O ≤ 500 mA	62	80		55	72		54	70		dB
		0°C ≤ Tj ≤ +125°C	62			55			54			dB
		V_{MIN} ≤ V_{IN} ≤ V_{MAX}	(8 ≤ V_{IN} ≤ 18)			(15 ≤ V_{IN} ≤ 25)			(18.5 ≤ V_{IN} ≤ 28.5)			V
R_O	Dropout Voltage	Tj = 25°C, I_{OUT} = 1A		2.0			2.0			2.0		V
	Output Resistance	f = 1 kHz		8			18			19		mΩ
	Short-Circuit Current	Tj = 25°C		2.1			1.5			1.2		A
	Peak Output Current	Tj = 25°C		2.4			2.4			2.4		A
	Average TC of V_{OUT}	0°C ≤ Tj ≤ +125°C, I_O = 5 mA		0.6			1.5			1.8		mV/°C
V_{IN}	Input Voltage Required to Maintain Line Regulation	Tj = 25°C, I_O ≤ 1A		7.5			14.6			17.7		V

Note 1: Thermal resistance of the TO-3 package (K, KC) is typically 4°C/W junction to case and 35°C/W case to ambient. Thermal resistance of the TO-220 package (T) is typically 4°C/W junction to case and 50°C/W case to ambient.

Note 2: All characteristics are measured with capacitor across the input of 0.22 µF, and a capacitor across the output of 0.1 µF. All characteristics except noise voltage and ripple rejection ratio are measured using pulse techniques (t_w ≤ 10 ms, duty cycle ≤ 5%). Output voltage changes due to changes in internal temperature must be taken into account separately.

Typical Performance Characteristics

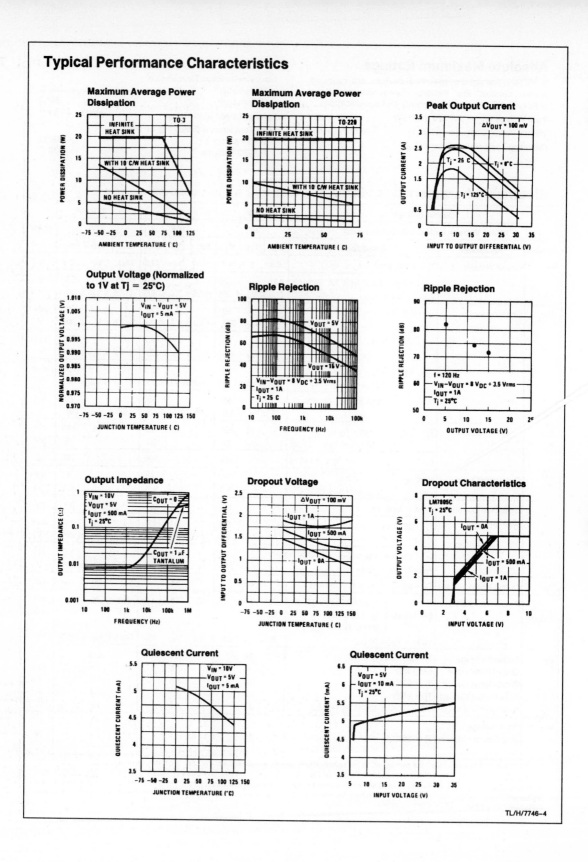

TL/H/7746–4